SpringerBriefs in Computer Science

For further volumes:
http://www.springer.com/series/10028

Springer Briefs in Computer Science

Duy Trong Ngo • Tho Le-Ngoc

Architectures of Small-Cell Networks and Interference Management

Duy Trong Ngo
School of Electrical Engineering
    and Computer Science
University of Newcastle
Callaghan, NSW, Australia

Tho Le-Ngoc
Department of Electrical
    and Computer Engineering
McGill University
Montreal, QC, Canada

ISSN 2191-5768   ISSN 2191-5776 (electronic)
ISBN 978-3-319-04821-5   ISBN 978-3-319-04822-2 (eBook)
DOI 10.1007/978-3-319-04822-2
Springer Cham Heidelberg New York Dordrecht London

Library of Congress Control Number: 2014932974

Printed on acid-free paper

Springer is part of Springer Science+Business Media (www.springer.com)

Preface

To accommodate the ever-increasing demand for mobile data, the wireless industry is facing with the urgent requirement of growing the capacity of mobile access networks by $1,000$ times. The extreme densification of small cells is currently the big hope to resolve the unprecedent "1000× data challenge" and to provide ubiquitous network coverage with an optimized grade of service. Small-cell heterogeneous networks represent a paradigm shift from the traditional centralized macrocell approach to a more self-organized solution, where small cells are deployed in conjunction with existing large cells at all possible venues, indoors and outdoors, and in all types and sizes. However, the coexistence of different types of network devices with diverse specifications on the same spectrum raises a new set of major design issues. These critical challenges urgently need to be solved to fully realize the promised benefits of small-cell solutions.

This SpringerBrief covers two important aspects of the emerging small-cell wireless heterogeneous networks. First, the architectures of small-cell networks are reviewed, with specific references to the current wireless network standards. Second, new adaptive power control and dynamic spectrum access techniques are discussed to promote a harmonized coexistence of diverse network entities in both 3G and 4G small-cell networks. Analytically devised from optimization and game theories, these autonomous solutions are shown to effectively manage the severe intra-tier and cross-tier interferences in small cells. The target audience of this informative and practical SpringerBrief is researchers and professionals working in wireless networking and interference management. The content is also valuable for advanced-level students interested in network communications and radio resource allocation.

We would like to acknowledge the financial supports from the Natural Sciences and Engineering Research Council of Canada and the Alexander Graham Bell Canada Graduate Scholarship.

Finally, we dedicate this work to our families.

Callaghan, NSW, Australia   Duy Trong Ngo
Montreal, QC, Canada   Tho Le-Ngoc

Contents

Acronyms

1-D	One-dimensional
2-D	Two-dimensional
3G	Third generation
3GPP	Third Generation Partnership Project
3GPP2	Third Generation Partnership Project 2
4G	Fourth generation
AGM	Arithmetic-geometric mean
AWGN	Additive white Gaussian noise
BER	Bit error rate
bps	Bit per second
BS	Base station
CDMA	Code-division multiple access
CINR	Channel-to-interference-plus-noise ratio
CN	Core Network
CoMP	Coordinated multipoint transmission and reception
CR	Cognitive radio
CSMA/CA	Carrier-sense multiple access with collision avoidance
D.C.	Difference-of-two-concave-functions
DSL	Digital subscriber line
EPC	Evolved Packet Core
FDMA	Frequency-division multiple access
FFT	Fast Fourier transform
FUE	Femtocell user equipment
GSM	Global System for Mobile Communications
HeNB	Home evolved Node B
HSPA	High Speed Packet Access
ICI	Intercell interference
IEEE	Institute of Electrical and Electronics Engineers
IP	Internet Protocol
IS-95	Interim Standard 95
KKT	Karush-Kuhn-Tucker

LTE	Long Term Evolution
MNO	Mobile network operator
MUE	Macrocell user equipment
NE	Nash equilibrium
NP	Non-deterministic polynomial-time
OAM	Operation, administration and management
OFDM	Orthogonal frequency-division multiplexing
OFDMA	Orthogonal frequency-division multiple access
PSD	Power spectral density
PU	Primary user
QoS	Quality of service
RNC	Radio Network Controller
Rx	Receiver
SC	Single-carrier
SCA	Successive convex approximation
SINR	Signal-to-interference-plus-noise ratio
SNR	Signal-to-noise ratio
SU	Secondary user
Tx	Transmitter
UE	User equipment
UMTS	Universal Mobile Telecommunications System
UTRAN	Universal Terrestrial Radio Access Network
VNI	Visual Networking Index
WCDMA	Wideband code-division multiple access
WiMAX	Wireless Interoperability for Microwave Access

Chapter 1
Dense Small-Cell Networks: Motivations and Issues

1.1 Mobile Data Traffic and Indoor Coverage Challenges

Globally, mobile data traffic has approximately doubled in each of the recent years and there are strong indications that this unprecedented trend will continue. According to the 2013 Ericsson Mobility Report, mobile data traffic has already surpassed voice traffic in 2009, and it is predicted to increase steadily whilst voice traffic only grows moderately [1]. Figure 1.1 shows that at the annual increase rate of 50%, the mobile traffic by the end of the year 2019 will be 10 times that of 2013. Moreover, in 2013 the traffic generated by mobile phones alone has exceeded that by all mobile PCs, mobile routers and tablets combined. Similarly, the 2013 Cisco Visual Networking Index (VNI) report has projected that the global mobile data traffic will go up at a compound annual growth rate (CAGR) of nearly 70% during the period 2012–2017 [2]. As seen from Fig. 1.2, a 13-time increase is expected by the end of 2017 with 11.2 exabytes generated per month.

Such an explosive traffic growth is a result of the increase in both the number of mobile users and the average amount of data information incurred by each user. This trend is fueled by the widespread adoption of wireless broadband, and further driven by larger-screen, smarter, faster and video-rich devices. The Cisco VNI report has also confirmed that video accounts for more than half of the total mobile data traffic, and by 2017 it will contribute to two-third of the global mobile data demand [2]. While video traffic is currently the main driver of such growth, machine-to-machine applications and connected vehicles and homes are expected to be the key contributors to the tremendous data traffic increase in a near future.

To accommodate the huge demand for mobile data in the coming years, the wireless industry is now facing with the real challenge of having to increase the capacity of mobile access networks by 1,000 times—the "1,000× challenge" [3]. Supported by all the current trends and future traffic predictions, the massive data growth can actually overwhelm the networks whose radio resources are limited.

D.T. Ngo and T. Le-Ngoc, *Architectures of Small-Cell Networks and Interference Management*, SpringerBriefs in Computer Science, DOI 10.1007/978-3-319-04822-2_1, © The Author(s) 2014

Fig. 1.1 Global mobile data traffic during 2011–2019 (Ericsson Mobility Report, 2013)

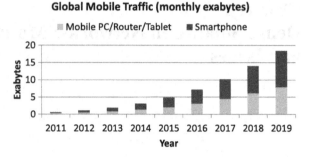

Fig. 1.2 Global mobile data traffic growth during 2012–2017 (Cisco VNI Mobile Forecast, 2013)

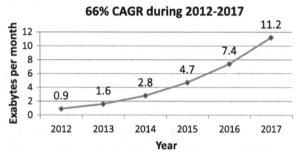

The finite radio spectrum is becoming scarcer, whereas technological advances are reaching their theoretical limits. New innovations are thus urgently required to tackle the $1,000\times$ data traffic challenge.

On the other hand, it is estimated that a majority of mobile traffic is originated and/or consumed from indoor environments. Mobile devices are increasingly used at home and workplace, even more often than when people are on the move. In any mobile access networks, buildings present a significant challenge because radio signals rapidly attenuate as they penetrate through the building walls. Since radio propagation is affected by the size, height and building materials, the traditional roof-top macro site approach has proved inefficient in providing adequate network coverage inside large buildings, especially in the dense urban terrain. New approaches are called for to offer ubiquitous coverage to this large population of indoor users. Such solutions will also offload a sizeable volume of data traffic from mobile networks to the indoor fixed networks, as shown in Fig. 1.3 [2], further contributing to the resolution of the $1,000\times$ data challenge.

1.2 Extreme Network Densification Solution

To meet the increasing demand for higher throughput and ubiquitous wireless coverage, several advanced solutions have been proposed in the literature [4, 5]. Allowing for concurrent use of different frequencies, carrier aggregation effectively

Fig. 1.3 Mobile traffic offloaded to indoor fixed networks during 2012–2017 (Cisco VNI Mobile Forecast, 2013)

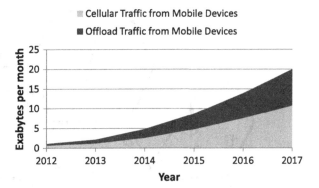

increases the bandwidth allocated to the UEs. Multiple-antenna solutions are attractive in that substantial diversity and multiplexing gains are exploited. As well, multiple cells employing coordinated multipoint (CoMP) techniques can coordinate their scheduling to serve UEs with unfavorable link conditions. CoMP techniques have been shown to be particularly useful in mitigating outage at the cell edges [6,7].

While significant technological innovations are urgently needed to successfully resolve the $1,000\times$ traffic challenge, advances on improving radio link performance have fast approaching the theoretical limits. Most likely, acquiring more radio spectrum and optimizing link efficiency are not sufficient to keep up with the exponential increase in traffic demand. The next capacity and performance leap beyond radio link improvements is now highly expected to come from a revolution in network topology, i.e., the hyper-dense deployment of small cells in conjunction with the existing large cells, at all possible venues indoors and outdoors, and in all types and sizes (femtocells, picocells, metrocells, remote radio heads, distributed antenna systems, etc.). It is strongly believed that the extreme cell densification is the most promising and scalable solution to meet the $1,000\times$ data increase [8–12].

1.2.1 Frequency Reuse Principle and Cellular Wireless Networks

Suppose that we are to provide full radio network coverage over a given geographical area. If one single transmitter is used, a large transmission power is required for the signals to reach all potential users in that area. In this case, only one transmission is allowed on any given radio frequency whilst there are limited radio frequencies available for transmission. Therefore, the total number of receivers that can be supported is restricted, making it impractical to serve a vast number of mobile wireless users.

To overcome such an inefficiency, wireless cellular networks based on the principle of frequency reuse have been proposed [13, 14]. A typical example of cellular networks is depicted in Fig. 1.4. Here, the area is divided into multiple

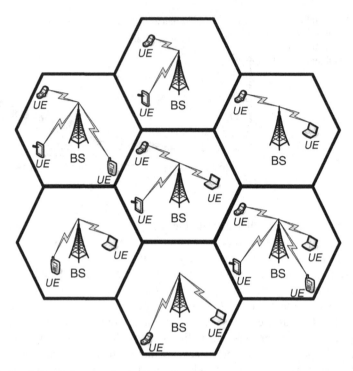

Fig. 1.4 A cellular wireless communication network

"cells," each of which has one base station (BS) that serves all the user equipments (UEs) operating within that cell. The downlink refers to the channel from the BS to UEs, whereas the uplink is the channel from UEs to the BS. Since the available radio frequencies are reused at each cell, the total number of supportable users is scaled with the number of deployed cells. Moreover, a high signal-to-noise ratio and better network coverage are made possible due to the reduced distance between a BS and its serviced UEs.

However, in this multicell and multiuser setting, the broadcast nature of the wireless medium results in the fundamental problem of signal interference. When several transmitters (i.e., BSs in the downlink and UEs in the uplink) emit their signals on the same frequency and within the same geographical location, the receiver (i.e., UEs in the downlink and BSs in the uplink) sensing that frequency may not be able to distinguish to which transmitter it is listening. Although the transmit power of a BS is limited to restrict the signal reception to the intended UEs within its cell, intercell interference still exists especially at the cell edges. The imperfect cell structure in practical scenarios also contributes to worsening the signal interference situations.

Fractional frequency reuse scheme can be used, where cells located close to one another are allotted with different frequency bands so as to avoid dominant intercell interference (ICI), i.e., cochannel interference. In contrast, cells that are sufficiently far from each other may reuse the same band. However, segmenting frequency reuse suffers from a reduced spectral efficiency. This problem is severe, given that radio spectrum is a scarce and expensive resource for telecommunication operators to acquire. To support the ever increasing demand by a huge number of wireless terminals for a higher quality-of-service (QoS), next-generation wireless networks are envisioned to employ a universal frequency reuse approach in which all cells operate on the same radio frequency. While offering potentially the most efficient spectral utilization, the universal reuse of radio spectrum may degrade network capacity if the critical issue of ICI is not properly addressed [11, 15].

1.2.2 Small-Cell Heterogeneous Network Deployment

The cellular network structure in Fig. 1.4 is homogeneous in the sense that the design specifications are similar for all cells (e.g., cell size, BS transmit power budget, allocated frequency). Generally designed to provide large coverage, homogeneous cellular networks are not efficient in offering high throughput, particularly in the very dense urban and indoor environments. Moreover, the current interference management approaches might not always work well in the low signal-to-interference-plus-noise ratio (SINR) regimes, in which transmit signals are substantially attenuated. This is especially true in the residential and office settings, wherein macrocell signals cannot reach indoor users due to the high level of isolation caused by building wall structures.

The extreme densification of small cells is recently proposed to realize the $1,000\times$ increase in network capacity and to provide ubiquitous network coverage. Also based on the principle of frequency reuse, small cells are deployed within the footprint area of the existing cells in all sizes and at all possible locations. As illustrated in Fig. 1.5, the resulting network structure is heterogeneous with small cells overlapping with the traditional large cells in both space and frequency. Without loss of generality, we refer to the traditional cells as macrocells and the small cells as femtocells in the rest of this brief.

Figure 1.6 depicts the coexistence of a macrocell and several femtocells in a wireless heterogeneous network. Specifically, macrocell user equipments (MUEs) establish links to their servicing macrocell BSs, while femtocell user equipments (FUEs) communicate with their respective femtocell BSs. These femtocell BSs are low-power, miniature wireless access points that are set at homes/offices and connected to backhaul networks via residential wireline broadband access links, e.g., digital subscriber lines (DSL), cable broadband connections, or optical fibers. Typically, the range of a femtocell is less than 50 m and it serves up to a dozen active users.

The benefits of dense femtocell deployments are summarized as follows [12, 16–18].

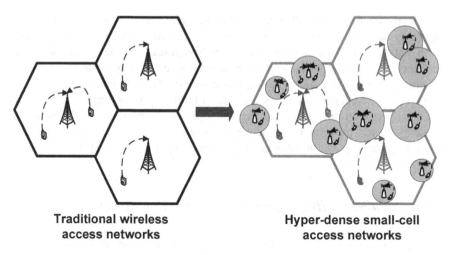

**Traditional wireless
access networks**

**Hyper-dense small-cell
access networks**

Fig. 1.5 From homogeneous cellular networks to heterogeneous small-cell networks

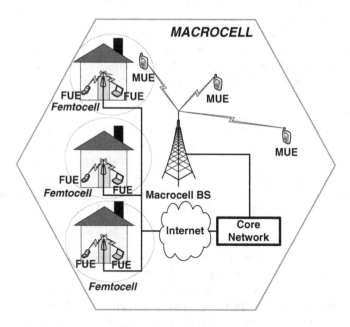

Fig. 1.6 A typical mixed macrocell/femtocell deployment scenario

- **Higher capacity:** With a large number of small cells, more users can be packed
  into a given area on the same radio spectrum, allowing for a greater area spectral
  efficiency (i.e., total number of active users per Hz per unit area). The low
  transmit power of femtocells and the signal isolation due to penetration losses
  provided by walls may also significantly limit the interference from neighboring

femtocells and macrocells. Often, this results in a higher femtocell capacity gain. Because macrocell no longer needs to transmit with high power to cover indoor areas, macrocell capacity grows too.

- **Better coverage with lower power consumption:** Thanks to the close proximity between FUEs and their serving femtocell BSs, these devices can lower their transmit power while still being able to achieve a high SINR. Ubiquitous network access is enabled with full network coverage even in the regions traditionally difficult to penetrate due to the shadowing effect. An improved coverage is crucial in the indoor areas where macrocell signals cannot reach.

- **Macrocell offload:** Traffic originating from indoor environments can be absorbed by femtocells via the IP backhauls, instead of being directed to the macrocell. Macrocell network can thus dedicate radio resources to better service its own users.

- **Cost effectiveness:** In the traditional cell-partitioning approach, a large number of expensive macrocell BSs are typically deployed after an extensive site survey and network planning process. On the contrary, femtocells can be easily integrated into an existing cellular network infrastructure. Mainly deployed by end users in a plug-and-play fashion, small-cell solution requires low capital expenditures and operating expenses, allowing for a cost-effective and scalable network evolution.

1.2.3   Technical Challenges in Small Cells

Small-cell heterogeneous networks represent a paradigm shift from the traditional centralized macrocell approach to a more uncoordinated and autonomous solution. Given that radio spectrum is limited, cochannel deployment is attractive in these heterogeneous networks, where FUEs share the same frequency bands with MUEs. With cochannel deployment, the peak data rates of legacy UEs are not impacted. Also, low-cost BSs are enabled while the higher-cost carrier aggregation-capable UEs are not required.

However, the coexistence of different types of network devices with diverse specifications on the same spectrum raises a new set of critical design issues that are inherent to heterogeneous networks. If the following technical challenges are not properly resolved, the benefits promised by femtocell deployments will be voided [5, 19].

- **Interference**: The interference situation is more acute and unpredictable in heterogeneous networks. This is because femtocells are randomly deployed without the network planning that would normally be undertaken. As femtocell BSs and FUEs can be moved or switched on/off at any time, the conventional network optimization turns out to be inefficient. In these cases, the operators cannot control the number and location of the newly deployed small cells. The rollout of many unplanned femtocells within the service area of regular macrocells also creates new cell boundaries. Accordingly, both MUEs and FUEs are more likely to suffer from strong ICI, especially those in the cell edges. In addition to the regular intra-tier interference within macrocells and femtocells, there are now cross-tier interferences from macrocells to femtocells and vice versa. The cross-tier interference is hard to control, and its effects can be particularly severe in many situations. In such cases, not only are the MUEs badly affected, a poor level of performance is achieved by the FUEs.

- **Service heterogeneity**: It remains unclear how to deliver an optimized QoS to the two distinct classes of users with different design specifications. As the spectrum owner, MUEs assume a higher priority in accessing the radio frequencies, and demand that their QoS not be degraded in spite of femtocell deployment. This implies that the cross-tier interference induced by FUEs to the macrocell network must be strictly controlled [20, 21]. On the other hand, the lower-tier FUEs wish to optimally configure their transmission parameters so as to exploit the residual network capacity, beyond what is needed to support the QoS requirements of all MUEs.

- **Limited backhaul**: The timely exchange of control and signaling information among femtocells and macrocells remains a major issue. Indeed, the residential network infrastructure connecting these cells can only support a limited network capacity for such a purpose. Since the wireline backhaul may not even belong to the network operators, it is possible that the communication experiences significant delays. The situation will become unacceptable if control and signaling information must be exchanged in a timely manner.

1.3 Structure of the Brief

This SpringerBrief covers two important aspects of small-cell wireless heterogeneous networks. First, we will review the architectures of small-cell networks, with specific references to the current wireless network standards. Second, we will present new results on distributed interference management for these emerging networks. We organize the rest of the brief as follows.

Chapter 2 provides an overview on the small-cell structures currently deployed in practical networks. We will begin with introducing the challenges and requirements of small-cell network architecture design. We will then review the various architectures available for small cells, with references to the 3GPP and 3GPP2 standards. In each architecture, key functional components and interfaces will be discussed. Next, the central issue of interference management in small-cell networks is presented, and the relevant state-of-the-art techniques in the literature for small-cell networks are also reviewed in detail.

In Chap. 3, we present new algorithms for distributed joint power and admission control in code-division multiple access (CDMA) based two-tier heterogeneous networks. For the upper-tier MUEs, we always maintain some prescribed minimum SINRs. For the lower-tier FUEs, we explicitly consider two different design objectives, namely, throughput-power tradeoff optimization and soft QoS provisioning. With an effective dynamic pricing scheme combined with admission control to indirectly manage the cross-tier interference, our proposed schemes mainly require local information to offer a maximized net utility of individual UEs.

Chapter 4 presents two Pareto-optimal power control algorithms for CDMA-based two-tier heterogeneous networks. Different from homogeneous network settings, the inevitable requirement of robustly protecting the QoS of all prioritized MUEs here lays a major obstacle that hinders the successful application of any available solutions. Directly targeting at this central issue, we propose the first algorithm that jointly maximizes the total utilities of both user classes. The second algorithm is applied to the scenario where only the sum utility of all FUEs needs to be maximized. We prove that both developed algorithms converge to their respective global optima, and more importantly, they can be implemented in a distributive manner at individual links. Effective mechanisms are also available to flexibly designate the access priority to MUEs and FUEs, as well as to fairly share radio resources among the UEs.

Chapter 5 proposes a joint subchannel and power allocation algorithm for the downlink of an orthogonal frequency-division multiple access (OFDMA) based mixed femtocell/macrocell network deployment. Specifically, the total throughput of all FUEs is maximized while the network capacity of an existing macrocell is always protected. We employ an iterative approach in which OFDM subchannels and BS transmit powers are alternatively assigned and optimized at every step. For a fixed power allocation, we prove that the optimal policy in each cell is to give subchannels to the UEs with the highest SINRs on those subchannels. For a given subchannel assignment, we adopt the successive convex approximation approach and devise three different power optimization solutions. We show that the joint subchannel and power allocation algorithm converges to the optimum of the original problem. While a central processing unit is required to implement the arithmetic-geometric mean approximation-based solution, each BS locally computes the optimal allocation for its own cell in the logarithmic and difference-of-two-concave-functions (D.C.) approximation-based solutions.

Chapter 6 presents joint subchannel assignment and power allocation algorithms that optimize the performance of an OFDMA-based cognitive femtocell network. Beside the interference constraints imposed at the macrocell network, we strictly enforce upper and lower bounds on the total number of subchannels granted to individual FUEs. This new requirement is particularly relevant in cognitive femtocell settings where the spectral activities of MUEs are highly dynamic, leaving a small opportunity for secondary access. We develop a dual decomposition framework for two criteria, namely, throughput maximization and power minimization, and devise distributed solutions. We show that the proposed algorithms achieve the actual global optimum with an affordable complexity in practical scenarios.

References

1. Ericsson, "Ericsson mobility report," Tech. Rep., Nov. 2013. [Online]. Available: http://www.ericsson.com/res/docs/2013/ericsson-mobility-report-november-2013.pdf
2. Cisco, "Cisco visual networking index: Global mobile data traffic forecast update 2012–2017," Tech. Rep., Feb. 2013. [Online]. Available: http://www.cisco.com/en/US/solutions/collateral/ns341/ns525/ns537/ns705/ns827/white_paper_c11-520862.pdf
3. 4G Americas, "Meeting the 1000x challenge: The need for spectrum, technology and policy innovation," Tech. Rep., Oct. 2013. [Online]. Available: http://www.4gamericas.org/documents/2013_4G%20Americas%20Meeting%20the%201000x%20Challenge%2010%204%2013_FINAL.pdf
4. M. Dottling, W. Mohr, and A. Osseiran, *Radio Technologies and Concepts for IMT-Advanced*. Wiley, Dec. 2009.
5. D. Lopez-Perez, I. Guvenc, G. de la Roche, M. Kountouris, T. Quek, and J. Zhang, "Enhanced intercell interference coordination challenges in heterogeneous networks," *IEEE Wireless Commun. Mag.*, vol. 18, no. 3, pp. 22–30, Jun. 2011.
6. M. Karakayali, G. Foschini, and R. Valenzuela, "Network coordination for spectrally efficient communications in cellular systems," *IEEE Wireless Commun.*, vol. 13, no. 4, pp. 56–61, Aug. 2006.
7. D. Gesbert, S. Hanly, H. Huang, S. S. Shitz, O. Simeone, and W. Yu, "Multi-cell MIMO cooperative networks: A new look at interference," *IEEE J. Select. Areas Commun.*, vol. 28, no. 9, pp. 1380–1408, Dec. 2010.
8. H. Claussen, "Performance of macro-and co-channel femtocells in a hierarchical cell structure," in *Proc. IEEE Intl. Symp. on Personal, Indoor and Mobile Radio Commun. (PIMRC)*, Sep. 2007, pp. 1–5.
9. I. Guvenc, M. R. Jeong, F. Watanabe, and H. Inamura, "A hybrid frequency assignment for femtocells and coverage area analysis for cochannel operation," *IEEE Commun. Lett.*, vol. 12, no. 12, pp. 880–882, Dec. 2008.
10. H. Claussen, L. T. W. Ho, and L. G. Samuel, "An overview of the femtocell concept," *Bell Labs. Tech. J.*, vol. 3, no. 1, pp. 221–245, May 2008.
11. D. Lopez-Perez, A. Valcarce, G. de la Roche, and J. Zhang, "OFDMA femtocells: A roadmap on interference avoidance," *IEEE Commun. Mag.*, vol. 47, no. 9, pp. 41–48, Sep. 2009.
12. J. Zhang and G. de la Roche, *Femtocells: Technologies and Deployment*, 1st ed. Wiley, 2010.
13. T. Rappaport, *Wireless Communications: Principles and Practice*, 2nd ed. Upper Saddle River, NJ, USA: Prentice Hall, 2001.
14. A. Goldsmith, *Wireless Communications*. Cambridge University, Aug. 2005.
15. R1-104968, "Summary of the description of candidate eICIC solutions, 3GPP Std." Madrid, Spain, Aug. 2010.
16. V. Chandrasekhar, J. G. Andrews, and A. Gatherer, "Femtocell networks: A survey," *IEEE Commun. Mag.*, vol. 46, no. 9, pp. 59–67, Sep. 2008.
17. D. Choi, P. Monajemi, S. Kang, and J. Villasenor, "Dealing with loud neighbors: The benefits and tradeoffs of adaptive femtocell access," in *Proc. IEEE Global Commun. Conf. (GLOBECOM)*, Dec. 2008, pp. 1–5.
18. C. Patel, M. Yavuz, and S. Nanda, "Femtocells [Industry Perspectives]," *IEEE Wirel. Commun.*, vol. 17, no. 5, pp. 6–7, Oct. 2010.
19. M. Yavuz, F. Meshkati, S. Nanda, A. Pokhariyal, N. Johnson, B. Raghothaman, and A. Richardson, "Interference management and performance analysis of UMTS/HSPA+ femtocells," *IEEE Commun. Mag.*, vol. 47, no. 9, pp. 102–109, Sep. 2009.
20. G. d. l. Roche, A. Valcarce, D. Lopez-Perez, and J. Zhang, "Access control mechanisms for femtocells," *IEEE Commun. Mag.*, vol. 48, no. 1, pp. 33–39, Jan. 2010.
21. S. Kishore, L. J. Greenstein, H. V. Poor, and S. C. Schwartz, "Uplink user capacity in a CDMA system with hotspot microcells: Effects of finite transmit power and dispersion," *IEEE Trans. Wireless Commun.*, vol. 5, no. 2, pp. 417–426, Feb. 2006.

Chapter 2
Architectures and Interference Management for Small-Cell Networks

2.1 Requirements and Reference Model for Small-Cell Network Architectures

The successful deployment of heterogeneous small-cell networks relies upon how one can integrate small cells into the existing mobile access networks to provide seamless device-to-core network connectivity. Defined for hierarchical deployments with network elements installed in secure premises, the existing mobile network architectures in GSM, UMTS, cdma2000 and LTE standards cannot be trivially extended to include small cells. In the ad hoc small-cell deployments, it is also particularly challenging to gain access to the dedicated high-performance links for interconnection and proprietary management systems, as is the case in the current architectures. New network structures are therefore needed to support small-cell integration with the following minimum requirements [1].

- *Scalability*: Whilst the current mobile networks only allow some few hundreds of macrocells to connect to the next level of the hierarchy, it is expected that small cells are massively deployed with many thousands of units per one single network. This calls for an architecture that can support sufficient scalability within the same network.
- *Transparent integration*: Small cells should be easily and transparently integrated into the existing mobile networks. At the same time, the additional load on the legacy infrastructure should be kept to the minimum.
- *Security*: Deployed at end-user premises, small cells typically operate in an insecure environment. As such, any proposed small-cell network architecture must guarantee a sufficient level of security for both mobile networks and end users.
- *Limited backhaul capacity*: The new network architecture must take into account the fact that small cells connect with one another via shared broadband IP links with variable performance. This situation is very different from that in the existing mobile networks where dedicated interconnection links are available.

D.T. Ngo and T. Le-Ngoc, *Architectures of Small-Cell Networks and Interference Management*, SpringerBriefs in Computer Science, DOI 10.1007/978-3-319-04822-2_2, © The Author(s) 2014

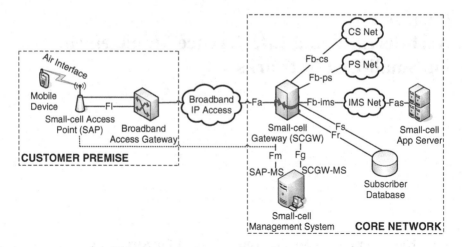

Fig. 2.1 Femto Forum reference network architecture [1–3]

Since the implementation details of small-cell architectures can vary considerably, it is important to have a consistent design approach to promote compatibility. Towards this end, the Femto Forum has provided a reference architecture for small cells that includes all the network elements and interfaces. This generic reference is applicable to a vast majority of network architectures and it can be used to compare alternative approaches. Illustrated in Fig. 2.1, the main functional components in this reference architecture are described as follows [1–3].

Small-cell access point: At the customer premise, the key component is a low-power hardware device called small-cell access point (SAP). Mobile users located inside the premise communicate with the SAP over the radio links, and typically up to a dozen of which can be supported by one SAP. The SAP connects to the core network via a broadband access gateway, which can either be a stand-alone device or be integrated in the SAP. The air interface between the SAP and mobile users can be single-carrier (e.g., CDMA) or multi-carrier (e.g., OFDMA). The Fl interface is used by the SAP to control the operating parameters in the broadband access gateway.

Broadband IP backhaul: As home base stations are equipped with more powerful processing capabilities, the traditional network protocol has essentially collapsed. At the same time, the Internet Protocol (IP) rapidly replaces the hierarchical telecommunications-specific transport protocols. It is proposed that small cells use flat networks, i.e., Internet-like, as the backhaul to transport data from home devices to the core network. The reference architecture employs broadband IP access links (e.g., digital subscriber lines, cables, fiber to the home) as the backhaul.

Small-cell gateway: The direct connectivity between the core network and the SAP is maintained by the small-cell gateway (SCGW). Together with signaling protocol and channel conversions, SCGW aggregates and integrates traffic from a large number of small cells into the existing mobile networks. The SCGW also implements security functions that authenticate and secure the connectivity with remote SAPs over the unsecured public broadband access links. The SCGW interfaces with the circuit-switch and packet-switch network segments of the mobile network operators (MNO) via Fb-cs and Fb-ps reference points, respectively. The SCGW–IMS network connectivity is supported by the Fb-ims interface.

With SCGW, the complexity and dimension of small-cell networks are hidden from the core network elements. It was earlier proposed that SAPs be kept simple and that all functions but radio be moved to SCGW. More recent solutions incline to support a flatter network by distributing much more functionalities to SAPs and keeping SCGW relatively simple. On the one side, SAPs support the front-end functions of Radio Network Controller (RNC), interact with end users, support mobility and perform radio resource management. On the other, SCGW supports back end RNC function, interfaces with core network and performs signalling aggregation. This approach allows for self-configured SAPs that support local services and local network access, enabling more cost-effective scalability.

Small-cell management system: Using Fm interface, the SAP management system (SAP-MS) can offer service provisioning and fault reporting of SAP devices. SAP-MS can handle tens of thousands of multi-vendor SAP units. Similarly, the SCGW management system (SCGW-MS) is expected to manage multiple SCGW devices via the Fg interface. The functions of SCGW-MS include traffic management, fault and alarm processing, and signaling protocol setting.

Subscriber database: The customer information such as SAP identity, network configurations and settings is stored in the subscriber databases. The SCGW accesses to these databases using the Fs and Fr interfaces.

2.2 Small-Cell Architectures in Wireless Network Standards

2.2.1 3GPP UMTS Small-Cell Architecture

The 3GPP Universal Mobile Telecommunications Systems (UMTS) consist of a Core Network (CN) and a Universal Terrestrial Radio Access Network (UTRAN). In particular, the UTRAN has a hierarchical architecture comprising RNCs and Node Bs, and it is connected to the CN via the Iu interface. As shown in Fig. 2.2, UMTS architecture is consistent with the generic model given in Fig. 2.1 albeit with the following modifications [2, 3]:

- Mobile device is now termed user equipment (UE),
- Small-cell access point is called home node B (HNB),
- Small-cell gateway is now HNB gateway (HNB-GW),
- Security gateway function is separated from HNB-GW,
- Fa interface is replaced by Iu-h interface.

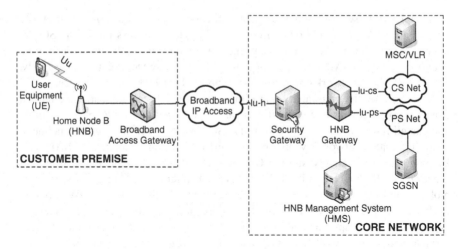

Fig. 2.2 3GPP UMTS small-cell network architecture [2, 3]

Deployed in the customer premise, the HNB is a low-power node that serves only one cell. The radio communication between the HNB and the UE is established via the Uu interface. In the core network, the HNB-GW plays the role of an RNC in that it concentrates multiple HNB connections on one side and connects to the MNO on the other side. While the connectivity between HNB-GW and HNBs is made possible with the Iuh interface, the HNB-GW employs Iu-cs and Iu-ps interfaces to connect with circuit-switch and packet-switch networks, respectively.

At the Iu-h reference point, a security gateway is deployed to protect the core network against security threats. Note that the security gateway can be implemented either as a separate physical element or be integrated to the HNB-GW. In this architecture, a new network element—HNB Management System (HMS)—is used to discover the HNB-GW, provide configuration data to HNBs, perform location verification of HNBs, etc.

It is worth noting the UMTS small-cell structure is able to offer architectural consistency. Since the HNB subsystem appears to core network as an existing Radio Network Subsystem (RNS), one can substantially reuse the existing network elements and protocols. At the same time, the HNB subsystem suffers from the existing limitations of RNS. Since a single HNB-GW can only address up to 65, 535 unique HNBs, handover from the regular macrocell to HNBs is not supported due to the limited cell addresses. Although hard handover from HNBs to macrocell is possible as macrocells can be unambiguously identified using the Cell Global Identification, soft handover from and to an HNB is not yet supported.

2.2.2 3GPP LTE Small-Cell Architecture

Evolved-UTRAN (E-UTRAN) is an evolution of the 3GPP UMTS radio access technology, where Long Term Evolution (LTE) is the radio interface and Evolved

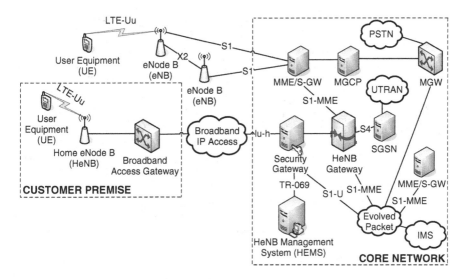

Fig. 2.3 3GPP LTE small-cell network architecture [2, 3]

Packet Core (EPC) is defined to accommodate the high-speed LTE access. The E-UTRAN consists of multiple evolved Node Bs (eNBs), which connect with one another via the X2 interface to support handover and with the EPC via the S1 interface for traffic and control purposes. Each eNB connects to the mobility management entity (MME) via the S1-MME interface and to the Serving Gateway (S-GW) via the S1-U interface.

Figure 2.3 shows that small cells can be integrated into the LTE structure with consistency. Compared with the reference model in Fig. 2.1, the following new definitions are introduced [2, 3]:

- Small-cell access point is now termed Home evolved NodeB (HeNB),
- Small-cell gate way is called HeNB gateway (HeNB GW),
- Security gateway function is separated from HeNB GW,
- Small-cell management system is called HeNB management system (HEMS).

The functions supported by the HeNB are identical to those by the eNB in the UMTS case [see Fig. 2.2]. Similarly, the procedures that run between the HeNB and the EPC are the same as between the eNB and the EPC. In this architecture, the HeNB GW is used to allow the S1 interface between the HeNB and the EPC, thereby supporting a large number of HeNBs. While the HeNB GW appears to the MME as an eNB, the former appears to the HeNB as the MME. Therefore, a HeNB is architecturally indistinguishable from an eNB in EPC. The handover support from a HeNB to an eNB and vice versa is available, whereas that among the HeNBs is still under investigation.

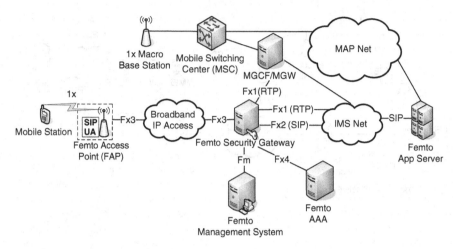

Fig. 2.4 CDMA2000 small-cell network architecture [2, 3]

2.2.3 3GPP2 CDMA2000 1x Small-Cell Architecture

The architecture for small-cell CDMA2000 1x deployment is shown in Fig. 2.4 [2, 3]. Different from the UMTS small-cell architecture, here the femto access point (FAP) includes a SIP user agent (SIP UA) to connect the 1x procedures on the mobile user side with the core network via a SIP/RTP interface. In this architecture, the femto security gateway (FSGW) maintains secure IP connectivity between the IMS core network and the FAP. On one side, an IPsec tunnel is established between the FSGW and the FAP via the Fx3 interface. On the other side, Fx1 interface transports RTP media packets to and from the FSGW, whereas the Fx2 interface implements the SIP signaling control.

The responsibility of the femto management system (FMS) includes configuring and managing the femtocell components via the newly-defined Fm interface. The femto AAA server authenticates the FAPs and shares security policy data with the FSGW. Using Fx4 interface, femto AAA server enables IPsec tunnels between the FAPs and the FSGW. Finally, the femto application server supports the interworking functions between the IMS network and the mobile carrier's MAP network.

2.2.4 Air Interfaces: CDMA vs. OFDMA

CDMA is used for medium access in UMTS, CDMA2000 and high speed packet access (HSPA) wireless standards. In a CDMA system, UEs in all cells are allowed to simultaneously transmit over all available frequency bands (see Fig. 2.5a).

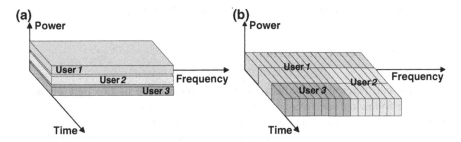

Fig. 2.5 Radio resource sharing in CDMA and OFDMA. (**a**) CDMA: All users share the same frequency at the same time (**b**) OFDMA: One subchannel is given to *at most* one user at a time

These transmissions are differentiated by the use of orthogonal codes, i.e., spreading codes, assigned to individual UEs. At the transmitting side, user's data signal is modulated with a spreading code to create a signal of a much larger bandwidth. At the receiving side, the cross-correlation of the received signal and the user's spreading code is calculated. When the resulting cross-correlation reaches its maximum, the corresponding data signal can be extracted. Since increasing the number of CDMA users only raises the noise floor in a linear manner, the system performance gradually degrades for all users. Hence, there is no absolute limit on the number of users that can be accommodated by the system.

On the other hand, frequency-selective fading is one of the major impairments of wireless channels, particularly in multipath environments such as indoor and urban areas. Since the channel responses differ among different frequencies, it can be challenging to alleviate the distortion that broadband signals experience when transmitted over such channels. In this situation, orthogonal frequency-division multiplexing (OFDM) signals are preferred because they are more robust to this type of fading.

The basic idea of OFDM is to divide the transmitted bitstream into many substreams, to be sent over a large number of closely-spaced orthogonal subchannels. Each subchannel is represented by one subcarrier, and one substream of data is transmitted through one subcarrier. Since individual subcarriers are modulated with a conventional modulation scheme at a much lower symbol rate, each of the resulting narrowband signals experiences frequency-flat fading. The IEEE Wireless Interoperability for Microwave Access (WiMAX) standard uses OFDM in the physical layer, whereas the 3GPP LTE standard employs OFDMA in the downlink and single-carrier FDMA (SC-FDMA) in the uplink [4]. Different from CDMA where each UE occupies all the spectrum at all time, a UE in OFDMA systems is allowed to only use a subgroup of OFDM subchannels, as shown in Fig. 2.5b.

2.3 Interference Management in Small-Cell Networks

2.3.1 Interference Scenarios

In a small-cell heterogeneous network, the communication of two tiers of users results the following interference scenarios. The intra-tier interference situation is similar to what occurs in homogeneous networks, where a macrocell interferes with other macrocells and a femtocell interferes with other femtocells. However, due to the significant difference in the transmit power limits, the most severe interference happens in the cross-tier scenario as illustrated in Fig. 2.6. In Scenario A, a victim cell-edge MUE is strongly interfered by the downlink transmission of a nearby femtocell BS. In Scenario B, an MUE located far away from its serving macrocell BS transmits at a high power in the uplink to compensate the path loss. This transmission creates strong interference to a nearby victim femtocell BS.

The severity of cross-tier interference also depends on the way that the radio frequency is allocated. In the orthogonal frequency allocation, distinct sets of frequencies are assigned to small-cell users (or femtocell users) and regular users (or macrocell users). Although the cross-tier interference can be completely avoided in this way, the resulting spectral efficiency is low because the radio spectrum is not efficiently reused. In the partially shared spectrum allocation option, macrocells have full access to the overall spectrum while femtocells are permitted to share a

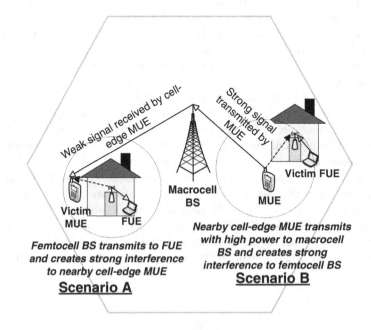

Fig. 2.6 Strong cross-tier interference in a mixed femtocell/macrocell deployment

subset of such spectrum. To mitigate the strong cross-tier interference, some radio channels are specifically reserved to only macrocells in the form of escape frequencies.

The highest degree of freedom is available in the universally shared spectrum allocation strategy, where both femtocell and macrocell users in all cells are allowed to utilize the same frequency bands. Potentially offering the most efficient use of the limited radio resources, this approach is highly promoted for next-generation wireless networks, and thus it will be assumed throughout this brief. However, the increased cross-tier interference in this case calls for more sophisticated schemes to mitigate the adverse effects of interference, thus fully realizing the potential gains of universal frequency reuse.

It is noteworthy that while CDMA systems provide resistance to narrowband interference, this property does not occur with broadband interference such as signals from other users. These signals remain as broadband interference even after the despreading process. With a unity spectral reuse factor where all UEs (either within the same cells or from different cells) share the same frequencies, interference is a critical problem in small-cell networks that is based on CDMA. With OFDMA being the air interface, intracell interference among UEs within the same cell can be suppressed. This is due to the assumption of exclusive subchannel assignment, i.e., one subchannel is used by at most one UE at a particular time (see Fig. 2.5b). However, aggressive frequency reuse allows a common spectrum to be shared among the UEs belonging to different cells. While interference averaging helps reduce the effect of interference in CDMA, it does not happen in OFDMA systems. Here, one interfering transmitter is enough to completely jam a given subchannel. It therefore remains challenging to effectively manage the ICI in OFDMA-based small-cell networks.

The successful rollout of small-cell wireless networks depends upon how the interference challenges are addressed. Optimized for the carefully-planned homogeneous networks, conventional approaches prove inefficient in managing the random and severe interference in small-cell scenarios. The stringent requirement of protecting macrocell performance imposes a new set of design constraints that may as well invalidate any available solutions. Moreover, the limited capacity for control and signaling also renders centralized mechanisms, which require the exchange of global network information, impractical in many situations.

2.3.2 Power Control for CDMA-Based Wireless Networks

2.3.2.1 Conventional Wireless Homogeneous Networks

Consider a CDMA-based multicell wireless homogeneous network. Let $p_i \geq 0$ be the transmit power of user i and σ_i be the power of the additive white Gaussian noise (AWGN). Denote the channel gain from the transmitter of user i to its receiver as $h_{i,i}$, and that from the transmitter of user j to the receiver of user $i \neq j$ as $h_{i,j}$.

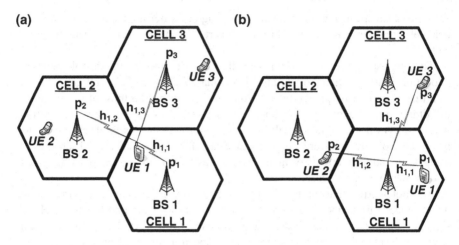

Fig. 2.7 Interference scenarios in a multicell homogeneous network. (**a**) Downlink (**b**) Uplink

Note that the "transmitter of user i" in the downlink is the BS that serves UE i, whereas in the uplink it is UE i. The received SINR of user i can be written as:

$$\gamma_i = \frac{h_{i,i}\, p_i}{\sum\limits_{j \neq i} h_{i,j}\, p_j + \sigma_i}. \tag{2.1}$$

As can be seen from (2.1), a large unwanted signal power $\sum_{j \neq i} h_{i,j} p_j$ may significantly decrease the SINR of user i, thereby degrading the quality of radio communication. Figure 2.7 illustrates two typical interference scenarios. In the downlink, UE 1 in cell 1 receives not only the intended signal from its serving BS 1 but also interfering signals from BSs 2 and 3. In the uplink, the signal transmitted by UE 1 to its BS 1 is interfered by those from UEs 2 and 3 in the two adjacent cells.

Power control has been proven to be very effective in dealing with interference in CDMA-based wireless networks. The most popular power control solution is probably the Foschini-Miljanic's algorithm [5], which enables users to eventually achieve their fixed target SINRs by iteratively adapting their transmit power according to:

$$p_i[t+1] = \frac{\gamma_i^{\min}}{\check{\gamma}_i[t]}\, p_i[t]. \tag{2.2}$$

Here, $\gamma_i^{\min}$ is the target SINR of user i, whereas $p_i[t]$ is the transmit power and $\check{\gamma}_i[t]$ is the measured SINR at the receiver of user i at time t.

It it worth noting that the simple algorithm in (2.2) can be implemented distributively by individual users, without requiring any form of network cooperation. As long as the target SINRs are feasible, (2.2) converges to a Pareto-optimal solution at a minimal aggregate transmit power $\sum_i p_i$. However, there is one major drawback in the Foschini-Miljanic's algorithm. If there exists an infeasible SINR target, the transmit power computed according to (2.2) will eventually diverge to infinity as each user i always attempts to meet its own required SINR at any cost. To deal with infeasible SINR targets, admission control algorithms are introduced in [6, 7].

The works in [8–11] investigate several other power control schemes from a game-theoretical point of view. The solutions devised from noncooperative games are appealing since they can be implemented in a decentralized fashion. In these games, individual users selfishly optimize their own performance, regardless of the actions of other users. Denote the utility (or payoff) function of user i as $U_i(p_i, \mathbf{p}_{-i})$, where $\mathbf{p}_{-i}$ is the power vector of all the users except i. The objective of each user i in the power-control game can be formally expressed as:

$$\max_{p_i \geq 0} U_i(p_i, \mathbf{p}_{-i}). \tag{2.3}$$

Depending on the type of utility function $U_i(\cdot)$, a number of games can be formulated whose solutions to the individual problem (2.3) exhibit different convergence properties. In most cases and under certain conditions, the underlying games settle at a Nash equilibrium (NE) $\mathbf{p}^* = [p_i^*]$, a stable and predictable state at which no user has any incentive to unilaterally change its transmit power level, i.e.,

$$U_i(p_i^*, \mathbf{p}_{-i}^*) \geq U_i(p_i, \mathbf{p}_{-i}^*), \quad \forall p_i \geq 0, \forall i. \tag{2.4}$$

Although the achieved NE gives a stable operating point, it is by no means guaranteed to be Pareto-efficient. To improve the efficiency of the equilibrium solutions, various pricing schemes are developed in [12, 13]. A pricing mechanism can implicitly enforce the cooperation among users while, at the same time, maintaining the noncooperative nature of the games. With pricing, the total utility of user i is:

$$U_{\text{tot},i}(p_i, \mathbf{p}_{-i}) = U_i(p_i, \mathbf{p}_{-i}) - C_i(p_i, \mathbf{p}_{-i}), \tag{2.5}$$

where $C_i(\cdot)$ denotes the cost imposed to user i. In each problem $\max_{p_i \geq 0} U_{\text{tot},i}$, various choices of utility and cost functions are available. Typically, the resulting solution is some modified version of the SINR balancing algorithm (2.2).

By selecting proper utilities and a linear cost $C_i(p_i, \mathbf{p}_{-i}) = a_i p_i$, [14, 15] show that noncooperative games with pricing can substantially enhance the NE if small deviations from the target SINRs are allowed. For instance, with $U_i(\gamma_i) = -(\gamma_i - \gamma_i^{\min})^2$ the transmit power can be updated according to [15]:

$$p_i[t+1] = \left(\frac{\gamma_i^{\min}}{\tilde{\gamma}_i[t]} p_i[t] - a_i \frac{p_i^2[t]}{\tilde{\gamma}_i^2[t]} \right)^+, \tag{2.6}$$

where $(\cdot)^+ = \max(\cdot, 0)$. Numerical results show that the enhanced Nash solution of [15] converges even faster than the SINR balancing algorithm in (2.2).

Still, it is unclear how far the Nash solutions given by [14, 15] are to the global optima of the power control problems. Using a different pricing scheme that is linearly proportional to SINR, i.e., $C_i(\gamma_i) = a_i \gamma_i$, [16] proves that the outcome of a noncooperative power control game in single-cell systems is a unique and Pareto-efficient NE. By setting dynamic prices for individual users and assuming noise-like ICI, various design goals can be met. In multicell communications where transmit powers of all users need to be jointly optimized across all cells, ICI cannot be simply treated as noise. The solutions by [16] are thus limited to single-cell scenarios.

Different from [5] where feasible SINR targets must be given a priori, [17] considers a decentralized joint optimization of SINR assignment and power allocation that is Pareto-optimal for multicell systems. It is argued that a fixed SINR assignment is not suitable for data-service networks, where target SINRs should instead be flexibly adjusted to the extent that the system capacity can still support. A high SINR is translated into better throughput and reliability, whereas a low SINR implies reduced data rates. In [17], a feasible SINR region is characterized in terms of the loads at BSs and the potential interference from UEs. With a re-parametrization via left Perron-Frobenius eigenvectors and a locally computable ascent direction, distributed Pareto-optimal solutions are derived for the uplink case.

2.3.2.2 Small-Cell Heterogeneous Networks

The results in [17] apply to homogeneous networks, in which there exist no differentiated classes of users with distinct access priority and design specifications. However, it is unclear how the proposed solutions account for the complicated coupling and strong interdependency among users in multi-tier heterogeneous networks. In such cases, the choices of target SINRs available to the lower-tier FUEs are much more limited. Also, strict QoS guarantees need to be enforced for the prioritized MUEs, and radio resources have to be dedicated to meet the demands of these users.

In the context of heterogeneous small-cell networks, [18–21] study various beamforming techniques to mitigate the undue cross-tier interference. Joint admission control and power management has also been examined in [22] for cognitive-CDMA networks. To protect the existing MUEs while enabling a scalable femtocell deployment, [23] proposes an uplink power control scheme for FUEs. Using open-loop and closed-loop techniques, this scheme adjusts the maximum transmit power as a function of the cross-tier interference level. Based on the actual interference at the macrocell BS, the proposed scheme can suppress the cross-tier interference. However, the devised solution is neither distributed nor Pareto-optimal.

For CDMA-based wireless heterogeneous networks, power control games are formulated and analyzed by [24, 25]. In particular, [25] considers the interference scenario depicted in Fig. 2.8, where p_i denotes the transmit power of the BS that serves UE i. Denoted as UE 0, the MUE is required to solve the following problem:

$$\max_{0 \le p_0 \le P^{\max}} U_0(p_0, \gamma_0 | \mathbf{p}_{-0}) = -(\gamma_0 - \gamma_0^{\min})^2. \tag{2.7}$$

Fig. 2.8 Downlink interference in a heterogeneous network

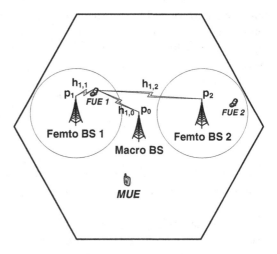

It is worth noting that the choice of utility function in (2.7) does not always guarantee the minimum SINR required by the MUE. Rather, only a "soft" SINR is provided. On the other hand, FUE i is to solve the following individual problem:

$$\max_{0 \le p_i \le P^{\max}} U_i(p_i, \gamma_i | \mathbf{p}_{-i}) = R(\gamma_i, \gamma_i^{\min}) + \bar{b}_i \frac{C(p_i)}{I_i(\mathbf{p}_{-i})}, \qquad (2.8)$$

with reward function $R(\cdot) = 1 - \exp\left[-\bar{a}_i(\gamma_i - \gamma_i^{\min})\right]$ and penalty function $C(\cdot) = -h_{0,i} p_i$. Here, $I_i(\cdot)$ is the interference power at the receiver of user i, and $\bar{a}_i, \bar{b}_i$ are constants. Note that because $C(\cdot)$ depends on the actual cross-tier interference $h_{0,i}$, explicit information about the cross-channel gains is required in the proposed algorithm. Due to the random fluctuations caused by shadowing and short-term fading effects, it can be quite challenging to estimate these channel values in practice.

2.3.3 Joint Subchannel-Power Allocation in OFDMA Networks

2.3.3.1 Conventional Wireless Homogeneous Networks

Compared to CDMA, OFDMA—the multiuser version of OFDM—provides three dimensions of diversity, i.e., time, frequency and users, for a more efficient allocation of radio resources. As there are multiple subchannels available in OFDMA, the resource optimization in this case faces another major technical difficulty, i.e., the subchannel assignment that allots radio frequencies to different UEs in multiple cells. To solve this combinatorial problem alone, direct search methods usually require a prohibitive computational complexity. Radio resource management for OFDMA-based networks relies upon efficient solutions that jointly optimize and assign powers and OFDM subchannels. Upon dividing the available spectrum into multiple subchannels, the SINR of UE k in cell m on subchannel n is expressed as:

$$\gamma_{m,k}^{(n)} = \frac{h_{m,k}^{(n)} p_m^{(n)}}{\sum_{s \neq m} h_{s,k}^{(n)} p_s^{(n)} + \sigma_k^{(n)}}, \tag{2.9}$$

where $p_m^{(n)}$ is the transmit power of BS m on subchannel n, $h_{m,k}^{(n)}$ the channel gain from BS m to UE k on subchannel n, and $\sigma_k^{(n)}$ the power of AWGN at the receiver of UE k on subchannel n.

Using noncooperative game theory, [26] solves the competition for radio resources in a multicell OFDMA-based network. Assuming that the interference from other UEs is fixed, the solution to the pure noncooperative game for individual UEs is of an iterative waterfilling type. In this case, it may happen that some undesirable NE with low performance is obtained or, even worse, there exists no NE at all. Moreover, if the cochannel interference is severe on some subchannels, the NE may not be optimal for the entire system. Motivated by this observation, [26] introduces the concept of a "virtual referee." By mandatorily changing of the game rules whenever needed, this referee can help improve the outcome of the formulated game. For example, it may reduce the transmit power of the UEs whose channel conditions are unfavorable. Those generating significant interference to other UEs may as well be prohibited from using certain subchannels. In doing so, the remaining cochannel UEs can share the corresponding subchannels in a more effective manner.

The study in [27] considers the problem of joint power allocation and subchannel assignment in the downlink of a multicell OFDMA network. Contrary to [26], the players in the formulated noncooperative game are the BSs, not the UEs. The players are responsible for allotting subchannels to the UEs within their cells, and deciding how much power to be distributed over those subchannels. Denote by $\mathbf{p} = [p_m^{(n)}]_{m,n} \succeq \mathbf{0}$ the network power vector that contains the transmit powers $p_m^{(n)}$ for all BSs m and all subchannels n. Also denote by $\boldsymbol{\rho}_m = [\rho_{m,k}^{(n)}]_{k,n}$ the channel assignment matrix of BS m, where $\rho_{m,k}^{(n)} = 1$ if subchannel n is assigned to UE k in cell m and $\rho_{m,k}^{(n)} = 0$ otherwise. The utility function of BS m is defined as:

$$U_m(\mathbf{p}, \boldsymbol{\rho}_m) = \sum_k \sum_n \rho_{m,k}^{(n)} \log \left(1 + \frac{p_m^{(n)} h_{m,k}^{(n)}}{\sum_{s \neq m} p_s^{(n)} h_{s,k}^{(n)} + \sigma_k^{(n)}} \right) - a_m \sum_n p_m^{(n)}, \tag{2.10}$$

where $a_m > 0$ is the price per unit of power. Given a network power vector **p**, it is shown that BS m assigns subchannel n to UE k^* if

$$k^* = k(m,n) = \arg\max_k \log \left(1 + \frac{p_m^{(n)} h_{m,k}^{(n)}}{\sum_{s \neq m} p_s^{(n)} h_{s,k}^{(n)} + \sigma_k^{(n)}} \right). \qquad (2.11)$$

Certainly, $\rho_{m,k^*}^{(n)}(\mathbf{p}) = 1$ in this case.

Once a fixed optimal subchannel assignment $\boldsymbol{\rho}_m^*$ is found, the optimal power allocation is derived as:

$$p_m^{(n)} = \left(\frac{1}{a_m + \lambda_m} - \frac{\sum_{s \neq m} p_s^{(n)} h_{s,k^*}^{(n)} + \sigma_{k^*}^{(n)}}{h_{m,k^*}^{(n)}} \right)^+, \qquad (2.12)$$

where $\lambda_m \left(\sum_n p_m^{(n)} - P^{\max} \right) = 0$ with $\lambda_m \geq 0$ being the Lagrange multiplier for the maximum total power constraint $P^{\max}$ at BS m. The allocations in (2.11) and (2.12) are performed iteratively until an equilibrium is finally reached. As proven in [27], such an iterative algorithm is guaranteed to converge to a unique NE under certain conditions. Usually, the stable operating points provided by the game-theoretical solutions do not globally maximize the network sum rates.

Different from [28–35] where the radio resources are allocated in a heuristic manner, [36] takes an optimization approach to solve the following problem of coordinated scheduling and power allocation in multicell OFDMA-based networks:

$$\max_{\mathbf{p}; \, \mathbf{k} = [k(m,n)]_{m,n}} \sum_m \sum_n w_{k(m,n)} r_{m,k(m,n)}^{(n)} \qquad (2.13)$$

$$\text{s.t. } \sum_n \mathbf{p}^{(n)} \preceq \mathbf{P}^{\max}.$$

Here, weight $w_{k(m,n)} \geq 0$ accounts for the priority of UE $k(m,n)$, $\mathbf{p}^{(n)} \succeq \mathbf{0}$ is the transmit power vector of all UEs on subchannel n, and $r_{m,k(m,n)}^{(n)} = \log \left(1 + \gamma_{m,k(m,n)}^{(n)}(\mathbf{p}^{(n)}) \right)$ is the corresponding throughput. The first proposed scheme—a multicarrier extension of the SCALE algorithm [37]—is proven to converge to a solution that satisfies the necessary optimality conditions of the nonconvex combinatorial problem (2.13). Using Lagrangian duality, the second scheme provides an optimal solution if the number of OFDM subchannels is very large [38, 39]. The third scheme is an improved iterative waterfilling algorithm, adapted to this multicell scenario. It is noted that all the solutions developed in [36] depend on a central unit to collect and process the complete channel state information. To alleviate the high complexity required by such solutions, [40] proposes a distributed low-complexity scheme based on the concept of a "reference user" to solve (2.13).

Considering the downlink of an OFDMA network, [41] addresses the problem of maximizing the weighted sum of the minimal UE rates of coordinated cells. In this case, the objective in (2.13) is modified as:

$$\sum_{\bar{m}} w_{\bar{m}} \min_{k \in \mathcal{K}_{\bar{m}}} \sum_{n} r_{\bar{m},k}^{(n)}, \tag{2.14}$$

where $w_{\bar{m}} \geq 0$ denotes the weight assigned to the smallest UE rate of cell $\bar{m}$, and $\mathcal{K}_{\bar{m}}$ the set of all UEs belonging to cell $\bar{m}$. Similar to [27], the centralized algorithm proposed by [41] alternatively optimizes the subchannel assignment and power allocation so that (2.14) keeps increasing until convergence. At each iteration, the allotment of subchannels is updated by resolving a mixed integer linear program for each cell. The optimal allocation of powers is found by a duality-based numerical algorithm. However, if a minimum rate constraint is strictly imposed to guarantee the QoS of some certain UE, the solutions in [36, 40, 41] are no longer applicable.

2.3.3.2 Small-Cell Heterogeneous Networks and Cognitive Femtocells

A joint subchannel and binary power allocation algorithm is developed in [42], where only one transmitter is allowed to send signals on each subchannel. Based on Lagrangian dual relaxation, [43, 44] propose various joint subchannel and power allocation schemes for OFDMA femtocells. It is assumed that the intra-tier inter-femtocell interference is negligible, whereas the cross-tier interference from the macrocell to femtocells is a constant. While these assumptions remarkably simplify the analysis, they are often not the case in practice. Moreover, network optimization for the existing macrocell is not considered at all in [43, 44].

In [45], the joint allocation of radio resource blocks and transmit powers is investigated for the downlink of OFDMA-based femtocells. The formulated exact-potential game is shown to always converge to an NE when the best-response adaptive strategy is applied [46]. Also taking a game-theoretical approach, [47] models macrocell BSs and femtocell BSs as the leaders and followers in a Stackelberg game [46]. In the hierarchical competition, a Stackelberg equilibrium, whose performance is better than that of an NE, is proven to exist under some mild conditions. As previously discussed, there is an ultimate need to protect the preferential MUEs in a mixed macrocell/femtocell network. This critical issue, however, has not been adequately addressed in [45, 47].

On the other hand, it has been confirmed that much of the licensed radio spectrum remains idle at any given time and location [48]. Spectrum utilization can thus be significantly improved by allowing (unlicensed) secondary users (SUs) to access spectrum holes unoccupied by (licensed) primary users (PUs). Cognitive radio [49–51] is promoted as an efficient technology to exploit the existence of spectrum portions unoccupied by PUs. While PUs still have a priority access to the radio spectrum, SUs are permitted to have a restricted access, subject to a constrained degradation on the PUs' QoS.

Spectrum pooling is an opportunistic access approach that enables public access to the already licensed frequency bands [52, 53]. The basic idea is to merge spectral ranges from different spectrum owners into a common pool, from which SUs may temporarily rent spectral resources during the idle periods of PUs. Here, the licensed system does not change while SUs access unused radio resources. In spectrum-pooling radio systems, OFDM is recognized as a highly promising candidate for SU transmission. This is mainly because of its flexibility in dynamically allocating the unused frequencies among SUs, and its ability to monitor PU spectral activities at no extra cost. However, OFDM transmission may cause mutual interference between PUs and SUs, due to the non-orthogonality of the respective signals [54, 55].

Several recent works propose that cognitive radio (CR) be used in heterogeneous small-cell networks, in that cognitive FUEs are allowed to opportunistically access the radio spectrum licensed to MUEs [56–58]. The roles of MUEs and FUEs in macrocell/femtocell settings correspond to those of PUs and SUs in CR networks, respectively. The existing results on radio resource management for OFDM-based CR networks can thus be applicable to two-tier cognitive femtocell networks.

In [59], an optimal power allocation scheme is devised to maximize the downlink capacity of a single SU, while guaranteeing that the interference induced to the PU is below a specified threshold. Similarly, [60] aims to maximize the CR link capacity, taking into account the availability of OFDM subchannels and the total interference limits at PUs. Extending the results in [59, 60] to multiuser scenarios, [61] aims at maximizing the discrete sum rate of a secondary network, constrained on the interference imposed to PU frequency bands. Subject to the per-subchannel power constraints (due to PU interference limits), [62] proposes a partitioned iterative water-filling algorithm that enhances the capacity of an OFDM CR system.

Zhang and Leung [63] attempts to solve the problem of resource allocation in multiuser OFDM-based CR systems. The main objective of [63] is to provide a satisfactory QoS to both real-time and non-real-time applications, despite the rapid variations in the available resources caused by the PUs' activities. In [64], the issue of downlink channel assignment and power control for FDMA-based cognitive networks has also been addressed, where BSs make opportunistic spectrum access to serve fixed-location UEs within their cells. Suboptimal schemes are derived to maximize the total number of supportable UEs, while guaranteeing the minimum SINR requirements of SUs and protecting the PUs.

To deal with the combinatorial OFDM subchannel assignment problem, the Lagrangian dual framework in [38] has proven to be especially useful. Considering networks with the coexistence of multiple primary and secondary links through OFDMA-based air-interface, [65] utilizes such an optimization framework to develop centralized and distributed algorithms. The design goal of [65] is to improve the total achievable sum rate of secondary networks, subject to interference constraints specified at PUs' receivers. Also based on Lagrangian duality, [66] studies the coexistence and optimization of a multicell CR network overlaid with a multicell primary network. The weighted sum rate of SUs over multiple cells is maximized in this case. For the downlink of a spectrum underlay OFDMA-based CR network, [67] proposes a joint subchannel-power allocation scheme that maximizes the CR network capacity. Here, the ICI among different CR cells is also controlled. With Lagrangian duality, the primal problem is decomposed into multiple dual subproblems, each of which is solved by an efficient algorithm. For Lagrangian dual framework to apply, the "frequency-sharing" condition must be strictly satisfied [38, 68].

References

1. S. Saunders, S. Carlaw, A. Giustina, R. R. Bhat, V. S. Rao, and R. Siegberg, *Femtocells: Opportunities and Challenges for Business and Technology*. Wiley, Jun. 2009.
2. J. Zhang and G. de la Roche, *Femtocells: Technologies and Deployment*, 1st ed. Wiley, 2010.
3. J. Boccuzzi and M. Ruggiero, *Femtocells: Design & Application*. McGraw Hill Education, Nov. 2010.
4. 4G Americas, *4G mobile broadband evolution: 3GPP Release 10 and beyond*, Feb. 2011.
5. G. J. Foschini and Z. Miljanic, "A simple distributed autonomous power control algorithm and its convergence," *IEEE Trans. Veh. Technol.*, vol. 42, no. 4, pp. 641–646, Nov. 1993.
6. M. Andersin, Z. Rosberg, and J. Zander, "Gradual removals in cellular PCS with constrained power control and noise," *Wireless Netw.*, vol. 2, pp. 27–43, 1996.
7. N. Bambos, S. C. Chen, and G. J. Pottie, "Channel access algorithms with active link protection for wireless communication networks with power control," *IEEE/ACM Trans. Netw.*, vol. 8, no. 5, pp. 583–597, Oct. 2000.
8. H. Ji and C.-Y. Huang, "Non-cooperative uplink power control in cellular radio systems," *Wireless Netw.*, vol. 4, no. 3, pp. 233–240, Jun. 1998.
9. Z. Han and K. J. R. Liu, "Noncooperative power-control game and throughput game over wireless networks," *IEEE Trans. Commun.*, vol. 53, no. 10, pp. 1625–1629, Oct. 2005.
10. J. W. Lee, R. R. Mazumdar, and N. B. Shroff, "Downlink power allocation for multi-class wireless systems," *IEEE/ACM Trans. Netw.*, vol. 13, no. 4, pp. 854–867, Aug. 2005.
11. E. Altman, T. Boulogne, R. El-Azouzi, T. Jiminez, and L. Wynter, "A survey of network games in telecommunications," *Comp. and Oper. Research*, pp. 286–311, Feb. 2006.
12. C. U. Saraydar, N. B. Mandayam, and D. J. Goodman, "Pricing and power control in a multicell wireless data network," *IEEE J. Select. Areas Commun.*, vol. 19, no. 10, pp. 1883–1892, Oct. 2001.
13. C. U. Saraydar, N. B. Mandayam, and D. J. Goodman, "Efficient power control via pricing in wireless data networks," *IEEE Trans. Commun.*, vol. 50, no. 2, pp. 291–303, Feb. 2002.
14. M. Xiao, N. B. Shroff, and E. K. P. Chong, "A utility-based power control scheme in wireless cellular systems," *IEEE/ACM Trans. Netw.*, vol. 11, no. 2, pp. 210–221, Apr. 2003.
15. S. Koskie and Z. Gajic, "A Nash game algorithm for SIR-based power control in 3G wireless CDMA networks," *IEEE/ACM Trans. Netw.*, vol. 13, no. 5, pp. 1017–1026, Oct. 2005.
16. M. Rasti, A. R. Sharafat, and B. Seyfe, "Pareto-efficient and goal-driven power control in wireless networks: A game-theoretic approach with a novel pricing scheme," *IEEE/ACM Trans. Netw.*, vol. 17, no. 2, pp. 556–569, Apr. 2009.
17. P. Hande, S. Rangan, M. Chiang, and X. Wu, "Distributed uplink power control for optimal SIR assignment in cellular data networks," *IEEE/ACM Trans. Netw.*, vol. 16, no. 6, pp. 1420–1433, Dec. 2008.
18. M. Husso, Z. Zheng, J. Hamalainen, and E. Mutafungwa, "Dominant interferer mitigation in closed femtocell deployment," in *Proc. IEEE Intl. Symp. Personal, Indoor and Mobile Radio Commun. Workshops*, Sep. 2010, pp. 169–174.

19. S. Ryoo, C. Joo, and S. Bahk, "Spectrum allocation with beamforming antenna in heterogeneous overlaying networks," in *Proc. IEEE Intl. Symp. on Personal, Indoor and Mobile Radio Commun. (PIMRC)*, Sep. 2010, pp. 1150–1155.
20. S. Park, W. Seo, Y. Kim, S. Lim, and D. Hong, "Beam subset selection strategy for interference reduction in two-tier femtocell networks," *IEEE Trans. Wireless Commun.*, vol. 9, no. 11, pp. 3440–3449, Nov. 2010.
21. S. Park, W. Seo, S. Choi, and D. Hong, "A beamforming codebook restriction for cross-tier interference coordination in two-tier femtocell networks," *IEEE Trans. Veh. Technol.*, vol. 60, no. 4, pp. 1651–1663, May 2011.
22. S. D. Roy, S. Mondal, and S. Kundu, "Performance of joint admission and power control algorithms in cognitive-CDMA network," in *Proc. Intl. Conf. on Comp. Commun. and Netw. Technologies (ICCCNT)*, Karur, India, Jul. 2010, pp. 1–6.
23. H.-S. Jo, C. Mun, J. Moon, and J.-G. Yook, "Interference mitigation using uplink power control for two-tier femtocell networks," *IEEE Trans. Wireless Commun.*, vol. 8, no. 10, pp. 4906–4910, Oct. 2009.
24. E. J. Hong, S. Y. Yun, and D.-H. Cho, "Decentralized power control scheme in femtocell networks: A game theoretic approach," in *Proc. IEEE Intl. Symp. on Personal, Indoor and Mobile Radio Commun. (PIMRC)*, Sep. 2009, pp. 415–419.
25. V. Chandrasekhar, J. G. Andrews, T. Muharemovic, and Z. Shen, "Power control in two-tier femtocell networks," *IEEE Trans. Wireless Commun.*, vol. 8, no. 8, pp. 4316–4328, Aug. 2009.
26. Z. Han, Z. Ji, and K. Liu, "Non-cooperative resource competition game by virtual referee in multi-cell OFDMA networks," *IEEE J. Select. Areas Commun.*, vol. 25, no. 6, pp. 1079–1090, Aug. 2007.
27. H. Kwon and B. G. Lee, "Distributed resource allocation through noncooperative game approach in multi-cell OFDMA systems," in *Proc. IEEE Intl. Conf. Commun. (ICC)*, vol. 9, Jun. 2006, pp. 4345–4350.
28. G. Li and H. Liu, "Downlink radio resource allocation for multi-cell OFDMA system," *IEEE Trans. Wireless Commun.*, vol. 5, no. 12, pp. 3451–3459, Dec. 2006.
29. H. Zhang, L. Venturino, N. Prasad, P. Li, S. Rangarajan, and X. Wang, "Weighted sum-rate maximization in multi-cell networks via coordinated scheduling and discrete power control," *IEEE J. Select. Areas Commun.*, vol. 29, no. 6, pp. 1214–1224, Jun. 2011.
30. K. Yang, N. Prasad, and X. Wang, "An auction approach to resource allocation in uplink OFDMA systems," *IEEE Trans. Signal Processing*, vol. 57, no. 11, pp. 4482–4496, Nov. 2009.
31. I. Koutsopoulos and L. Tassiulas, "Cross-layer adaptive techniques for throughput enhancement in wireless OFDM-based networks," *IEEE/ACM Trans. Netw.*, vol. 14, no. 5, pp. 1056–1066, Oct. 2006.
32. K. Yang, N. Prasad, and X. Wang, "A message-passing approach to distributed resource allocation in uplink DFT-spread-OFDMA systems," *IEEE Trans. Commun.*, vol. 59, no. 4, pp. 1099–1113, Apr. 2011.
33. N. Ksairi, P. Bianchi, P. Ciblat, and W. Hachem, "Resource allocation for downlink cellular OFDMA systems - Part I: Optimal allocation," *IEEE Trans. Signal Processing*, vol. 58, no. 2, pp. 720–734, Feb. 2010.
34. M. Pischella and J.-C. Belfiore, "Weighted sum throughput maximization in multicell OFDMA networks," *IEEE Trans. Veh. Technol.*, vol. 59, no. 2, pp. 896–905, Feb. 2010.
35. B. Da and R. Zhang, "Cooperative interference control for spectrum sharing in OFDMA cellular systems," in *Proc. IEEE Intl. Conf. Commun. (ICC)*, Jun. 2011, pp. 1–5.
36. L. Venturino, N. Prasad, and X. Wang, "Coordinated scheduling and power allocation in downlink multicell OFDMA networks," *IEEE Trans. Veh. Technol.*, vol. 58, no. 6, pp. 2835–2848, Jul. 2009.
37. J. Papandriopoulos and J. S. Evans, "SCALE: A low-complexity distributed protocol for spectrum balancing in multiuser DSL networks," *IEEE Trans. Inform. Theory*, vol. 55, no. 8, pp. 3711–3724, Aug. 2009.
38. W. Yu and R. Lui, "Dual methods for nonconvex spectrum optimization of multicarrier systems," *IEEE Trans. Commun.*, vol. 54, no. 7, pp. 1310–1322, Jul. 2006.

39. R. Cendrillon, W. Yu, M. Moonen, J. Verlinden, and T. Bostoen, "Optimal multiuser spectrum balancing for digital subscriber lines," *IEEE Trans. Commun.*, vol. 54, no. 5, pp. 922–933, May 2006.

40. K. Son, S. Lee, Y. Yi, and S. Chong, "REFIM: A practical interference management in heterogeneous wireless access networks," *IEEE J. Select. Areas Commun.*, vol. 29, no. 6, pp. 1260–1272, Jun. 2011.

41. T. Wang and L. Vandendorpe, "Iterative resource allocation for maximizing weighted sum min-rate in downlink cellular OFDMA systems," *IEEE Trans. Signal Processing*, vol. 59, no. 1, pp. 223–234, Jan. 2011.

42. J. Kim and D.-H. Cho, "A joint power and subchannel allocation scheme maximizing system capacity in indoor dense mobile communication systems," *IEEE Trans. Veh. Technol.*, vol. 59, no. 9, pp. 4340–4353, Nov. 2010.

43. H. Zhang, W. Zheng, X. Chu, X. Wen, M. Tao, A. Nallanathan, and D. Lopez-Perez, "Joint subchannel and power allocation in interference-limited OFDMA femtocells with heterogeneous QoS guarantee," in *Proc. IEEE Global Commun. Conf. (GLOBECOM)*, Dec. 2012, pp. 4794–4799.

44. L. Li, C. Xu, and M. Tao, "Resource allocation in open access OFDMA femtocell networks," *IEEE Wireless Commun. Lett.*, vol. 1, no. 6, pp. 625–628, Dec. 2012.

45. L. Giupponi and C. Ibars, "Distributed interference control in OFDMA-based femtocells," in *Proc. IEEE Intl. Symp. on Personal, Indoor and Mobile Radio Commun. (PIMRC)*, Sep. 2010, pp. 1201–1206.

46. D. Fudenberg and J. Tirole, *Game Theory*. MIT Press, 1991.

47. S. Guruacharya, D. Niyato, D. I. Kim, and E. Hossain, "Hierarchical competition for downlink power allocation in OFDMA femtocell networks," *IEEE Trans. Wireless Commun.*, vol. 12, no. 4, pp. 1543–1553, Apr. 2013.

48. FCC Spectrum Policy Task Force, "Report of the spectrum efficiency working group," Federal Communications Commission, Tech. Rep. ET Docket No. 02-135, Nov. 2002.

49. J. Mitola and G. Q. Maguire, "Cognitive radio: Making software radios more personal," *IEEE Personal Commun. Mag.*, vol. 6, no. 4, pp. 13–18, Aug. 1999.

50. S. Haykin, "Cognitive radio: Brain-empowered wireless communications," *IEEE J. Select. Areas Commun.*, vol. 23, no. 2, pp. 201–220, Feb. 2005.

51. Q. Zhao and B. M. Sadler, "A survey of dynamic spectrum access," *IEEE Signal Processing Mag.*, pp. 79–89, May 2007.

52. T. Weiss and F. Jondral, "Spectrum pooling: An innovative strategy for the enhancement of spectrum efficiency," *IEEE Commun. Mag.*, vol. 42, no. 3, pp. S8–14, Mar. 2004.

53. U. Berthold, F. Jondral, S. Brandes, and M. Schnell, "OFDM-based overlay systems: A promising approach for enhancing spectral efficiency [Topics in Radio Communications]," *IEEE Commun. Mag.*, vol. 45, no. 12, pp. 52–58, Dec. 2007.

54. T. Weiss, J. Hillenbrand, A. Krohn, and F. Jondral, "Mutual interference in OFDM-based spectrum pooling systems," in *Proc. IEEE Vehicular Technology Conf. (VTC)*, vol. 4, May 2004, pp. 1873–1877.

55. B. Farhang-Boroujeny and R. Kempter, "Multicarrier communication techniques for spectrum sensing and communication in cognitive radios," *IEEE Commun. Mag.*, pp. 80–85, Apr. 2008.

56. G. Gur, S. Bayhan, and F. Alagoz, "Cognitive femtocell networks: An overlay architecture for localized dynamic spectrum access [Dynamic Spectrum Management]," *IEEE Wirel. Commun.*, vol. 17, no. 4, pp. 62–70, 2010.

57. S. Al-Rubaye, A. Al-Dulaimi, and J. Cosmas, "Cognitive femtocell," *IEEE Veh. Technol. Mag.*, vol. 6, no. 1, pp. 44–51, 2011.

58. R. Xie, F. Yu, H. Ji, and Y. Li, "Energy-efficient resource allocation for heterogeneous cognitive radio networks with femtocells," *IEEE Trans. Wireless Commun.*, vol. 11, no. 11, pp. 3910–3920, 2012.

59. G. Bansal, M. Hossain, and V. Bhargava, "Optimal and suboptimal power allocation schemes for OFDM-based cognitive radio systems," *IEEE Trans. Wireless Commun.*, vol. 7, no. 11, pp. 4710–4718, Nov. 2008.

60. Z. Hasan, G. Bansal, E. Hossain, and V. Bhargava, "Energy-efficient power allocation in OFDM-based cognitive radio systems: A risk-return model," *IEEE Trans. Wireless Commun.*, vol. 8, no. 12, pp. 6078–6088, Dec. 2009.

61. T. Qin and C. Leung, "Fair adaptive resource allocation for multiuser OFDM cognitive radio systems," in *Proc. Second Intl. Conf. Communications and Networking in China (CHINACOM)*, Aug. 2007, pp. 115–119.

62. P. Wang, M. Zhao, L. Xiao, S. Zhou, and J. Wang, "Power allocation in OFDM-based cognitive radio systems," in *Proc. IEEE Global Telecommun. Conf. (GLOBECOM)*, Dec. 2007, pp. 4061–4065.

63. Y. Zhang and C. Leung, "Cross-layer resource allocation for mixed services in multiuser OFDM-based cognitive radio systems," *IEEE Trans. Veh. Technol.*, vol. 58, no. 8, pp. 4605–4619, Oct. 2009.

64. A. T. Hoang and Y.-C. Liang, "Downlink channel assignment and power control for cognitive networks," *IEEE Trans. Wireless Commun.*, vol. 7, no. 8, pp. 3106–3117, Aug. 2008.

65. P. Cheng, Z. Zhang, H.-H. Chen, and P. Qiu, "Optimal distributed joint frequency, rate and power allocation in cognitive OFDMA systems," *IET Communications*, vol. 2, no. 6, pp. 815–826, Jul. 2008.

66. Y. Ma, D. I. Kim, and Z. Wu, "Optimization of OFDMA-based cellular cognitive radio networks," *IEEE Trans. Commun.*, vol. 58, no. 8, pp. 2265–2276, Aug. 2010.

67. K. W. Choi, E. Hossain, and D. I. Kim, "Downlink subchannel and power allocation in multi-cell OFDMA cognitive radio networks," *IEEE Trans. Wireless Commun.*, vol. 10, no. 7, pp. 2259–2271, Jul. 2011.

68. D. P. Bertsekas, *Nonlinear Programming*, 2nd ed. Boston: Athena Scientific, 1999.

Chapter 3
Distributed Interference Management in Heterogeneous CDMA Small-Cell Networks

This chapter presents joint power and admission control algorithms for two-tier CDMA-based heterogeneous networks [1, 2]. The fundamental difference between the setting considered here and that in traditional CDMA wireless networks is the differentiated classes of users with distinct access priorities and design requirements. The prioritized MUEs demand that their QoS requirements are always maintained in the first place, whereas the lower-tier FUEs attempt to optimize their performance by exploiting the remaining system resources. Specifically, two practical scenarios are investigated: (1) FUEs desire to balance their achieved throughput with the corresponding power expenditure, and (2) FUEs demand certain "soft" QoS requirements, expressed in terms of minimum attained SINRs. In lightly-loaded networks, an effective mechanism is also proposed that helps to better utilize the network capacity and improve the performance of MUEs. Convergence properties of the proposed algorithms are rigorously analyzed and potential extensions are presented to further emphasize the attractiveness of the developed solutions.

It is noteworthy that, whilst closest in spirit with [3], the work presented here distinguishes itself in at least two key aspects. Firstly, to represent the net utility of FUEs, [3] uses a penalty function that depends on the actual cross-tier interference, and hence requiring explicit information about the cross-channel gains. On the contrary, this work proposes an effective dynamic pricing scheme combined with admission control to indirectly manage the cross-tier interference. Together with their distributive nature, the developed schemes are more tractable in view of practical implementation under the limited backhaul network capacity available for femtocells. Secondly, the choice of utility function for MUEs in [3] does not always guarantee that the minimum required SINRs are achieved for these prioritized users. By selecting a sigmoid function to represent the macrocell utility, the devised joint power and admission control algorithms are capable of robustly protecting the performance of all active MUEs.

D.T. Ngo and T. Le-Ngoc, *Architectures of Small-Cell Networks and Interference Management*, SpringerBriefs in Computer Science, DOI 10.1007/978-3-319-04822-2_3,

3.1   System Model and Assumptions

We consider a two-tier wireless network with power-controlled UEs. Specifically, we investigate the scenario where a macrocell serving M MUEs is underlaid with M_f femtocells. Although we assume CDMA for multiple access, the results obtained in this chapter are applicable to single-carrier wireless systems in general. Assume that femtocell i has K_i FUEs and define $K = \sum_{i=1}^{M_f} K_i$ as the total number of FUEs in the network. Further assume that the association of FUEs with their closest femtocell BSs is fixed during the runtime of the power and admission control processes. Denote the set of MUEs and FUEs by $\mathscr{L}_m$ and $\mathscr{L}_f$, respectively. The set of all UEs is then $\mathscr{L} = \mathscr{L}_m \cup \mathscr{L}_f$. An example of such a network is illustrated in Fig. 3.1.

The results obtained in this work are applicable to *both downlink and uplink* scenarios. By "the transmitter of UE i" we refer to the BS that serves wireless terminal $i \in \mathscr{L}$ in the downlink case, whereas in the uplink case it is the wireless terminal i. We consider a snapshot model where the channel gains remain unchanged during the runtime of the power and admission control algorithms. Let p_i be the transmit power of UE i and σ_i the power of additive white Gaussian noise measured in the spectrum bandwidth at the receiving end of UE $i \in \mathscr{L}$. Denote the channel gain from the transmitter of UE i to its receiver by $h'_{i,i}$, and that from the transmitter of UE j to the receiver of UE $i \neq j$ by $h_{i,j}$. The received SINR of UE $i \in \mathscr{L}$ is:

$$\gamma_i = \frac{G h'_{i,i} p_i}{\sum_{j \neq i} h_{i,j} p_j + \sigma_i}, \qquad (3.1)$$

where G is the system processing gain.

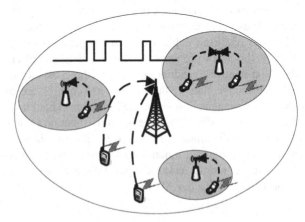

Fig. 3.1 Example of a two-tier CDMA wireless heterogeneous network (*Tier 1*: a macrocell; *Tier 2*: several femtocells)

Note that the first term in the denominator of (3.1) includes both intra-cell and cross-tier interferences, i.e., *aggregated* interference from all MUEs and FUEs except the considered UE i (which can be either an MUE or an FUE). In the downlink case, the channel gain $h_{i,j}$ reduces to $h'_{i,i}$ for the intra-cell interference, while $h_{i,j}$ is termed the *cross-channel* gain for the cross-tier interference. For notational convenience, let $h_{i,i} = Gh'_{i,i}$ where the processing gain G is absorbed into the channel gain $h'_{i,i}$. The received SINR of UE $i \in \mathcal{L}$ can then be expressed as:

$$\gamma_i = \frac{h_{i,i}\, p_i}{\sum_{j \neq i} h_{i,j}\, p_j + \sigma_i}.$$

(3.2)

In cellular wireless networks such as IS-95, WCDMA and LTE networks, regardless of traffic types, a minimum SINR is required at the receiver for a minimum data rate to be supported. While the maintenance of such minimum SINR targets is well-justified for voice users to achieve a certain desired bit error rate (BER), it is also applicable to data users, especially those with delay-sensitive applications. In our modeling framework, different SINR thresholds are assigned to different UEs, depending on their access priority and application requirements. Given a desired threshold $\gamma_i^{\min}$, we assume that the prioritized MUE $i \in \mathcal{L}_m$ requires that:

$$\gamma_i \geq \gamma_i^{\min}.$$

(3.3)

On the other hand, each FUE $i \in \mathcal{L}_f$, which is of a lower access priority, is assumed to suppress transmission whenever its attained SINR falls below a predefined threshold $\underline{\gamma}_i$. The rationale behind this assumption is that a negligible level of SINR would not help anything at all, but only create unnecessary interference to other UEs. Therefore, we require that an active FUE $i \in \mathcal{L}_f$ must have that:

$$\gamma_i \geq \underline{\gamma}_i.$$

(3.4)

We employ a utility function $U_i(\gamma_i)$ and a cost function $C_i(p_i)$ to represent the degree of satisfaction of UE $i \in \mathcal{L}$ to the service quality and the cost incurred when performing such a service, respectively. $U_i(\gamma_i)$ takes larger values with preferred services, while a high cost $C_i(p_i)$ can be used to control the selfish behaviors of user i. It is the interest of UE $i \in \mathcal{L}$ to maximize its own net utility defined as:

$$U_{\text{tot},i} = U_i(\gamma_i) - C_i(p_i).$$

(3.5)

In fact, (3.5) is a standard way to define the payoff function for network entities (i.e., wireless UEs and BSs). Given the transmit power of other UEs, the net utility can be maximized by power adaptation dynamically performed at individual links.

Assume that $U_i(\gamma_i)$ is strictly concave in γ_i whereas $C(p_i)$ is convex in p_i. The necessary condition for the optimality of (3.5) can be obtained by taking the derivative of $U_{\text{tot},i}$, which is also strictly concave in p_i, and equating to zero as follows.

$$\frac{dU_{\text{tot},i}}{dp_i} = \frac{dU_i}{d\gamma_i}\frac{d\gamma_i}{dp_i} - \frac{dC_i}{dp_i} = 0. \tag{3.6}$$

Denote the derivatives of $U_i(\gamma_i)$ and $C_i(p_i)$ as $U_i'(\gamma_i)$ and $C_i'(p_i)$, respectively. Upon noting that $\dfrac{d\gamma_i}{dp_i} = \dfrac{h_{i,i}}{I_i} = \dfrac{\gamma_i}{p_i}$, we have that:

$$U_i'(\gamma_i) = \frac{p_i}{\gamma_i}C_i'(p_i) = \frac{I_i}{h_{i,i}}C_i'(p_i), \tag{3.7}$$

where $I_i = \sum_{j \neq i} h_{i,j} p_j + \sigma_i$ is the total noise and interference power at the receiving side of UE $i \in \mathscr{L}$. From (3.7), the optimal target SINR can be derived as:

$$\hat{\gamma}_i = f_i^{-1}\left(\frac{I_i}{h_{i,i}}C_i'(p_i)\right), \tag{3.8}$$

with $f_i(\gamma_i) = U_i'(\gamma_i)$ in the *concave part* of $U_i(\gamma_i)$ where a local maximum is possible. Based on $\hat{\gamma}_i$ in (3.8), the following iterative power-update rule can be applied [4]:

$$p_i[t+1] = \hat{\gamma}_i[t]\frac{I_i[t]}{h_{i,i}} = \frac{\hat{\gamma}_i[t]}{\gamma_i[t]}p_i[t], \tag{3.9}$$

where $\gamma_i[t]$ is the actual SINR of UE i at iteration t. In fact, (3.9) represents a more general power-control rule compared with the following well-known update [5, 6]:

$$p_i[t+1] = \frac{\gamma_i^{\min}}{\gamma_i[t]}p_i[t]. \tag{3.10}$$

Specifically, the minimum required SINR $\gamma_i^{\min}$ on the right-hand side of (3.10) is replaced by the adaptive SINR threshold $\hat{\gamma}_i[t]$ in (3.8).

In what follows, we will show how to choose functions $U_i(\gamma_i)$ and $C_i(p_i)$, together with their operating parameters, to design efficient distributed power and admission control algorithms for both MUEs and FUEs. The key aspect that makes the existing algorithms (such as those presented in [7, 8]) unsuitable for our current purpose is that the minimum SINRs of the prioritized MUEs should be maintained at all times. As a direct consequence, FUEs must have their transmit powers properly controlled or, if needed, may even be removed for the sake of protecting MUEs.

3.2 Distributed Joint Power and Admission Control Algorithms

3.2.1 QoS Guarantee for MUEs

In the design of their power control scheme, [4] recommends the use of a sigmoid utility function and a linear cost function. For our problem at hand, by employing similar utility and cost functions for the MUEs and via properly tuning their control parameters, we can develop an efficient power control algorithm that is capable of maintaining the minimum SINR requirements for these UEs. Specifically, we select the following utility and cost functions for MUE $i \in \mathcal{L}_m$:

$$U_i(\gamma_i) = \frac{1}{1 + \exp[-b_i(\gamma_i - c_i)]}, \tag{3.11}$$

$$C_i(p_i) = a_i^{(m)} p_i. \tag{3.12}$$

Here, $b_i > 0$ and c_i, respectively, control the steepness and the center of the sigmoid function, whereas $a_i^{(m)}$ is the pricing coefficient.

Function $U_i(\gamma_i)$ in (3.11) naturally captures the value of the service provided to UE i. By noting that $U_i(0) \approx 0$ (as we can made $e^{b_i c_i}$ very large), $U_i(\infty) = 1$, and that $U_i(\gamma_i)$ is increasing with respect to γ_i, it is clear that UE i is increasingly satisfied as the quality of the offered service, expressed in terms of the achieved SINR γ_i, improves. On the other hand, transmit power is a valuable system resource. The linear cost in (3.12) is chosen to reflect the expenses of power consumption to UE i, while allowing the simplicity of subsequent analysis. As will be shown later, the use of dynamic values of $a_i^{(m)}$ may significantly affect the resulting equilibrium of the developed algorithms.

Importantly, the choice of sigmoid function enables the design of efficient schemes that guarantee the minimum SINRs imposed by MUEs. With (3.11) and (3.12), (3.7) becomes:

$$U_i'(\gamma_i) = f_i(\gamma_i) = \frac{a_i^{(m)} I_i}{h_{i,i}}. \tag{3.13}$$

From this relationship, it is straightforward to see that the optimal SINR target is:

$$\hat{\gamma}_i = f_i^{-1}\left(\frac{a_i^{(m)} I_i}{h_{i,i}}\right). \tag{3.14}$$

Since $f_i(\cdot)$ is defined in the concave part of $U_i(\gamma_i)$, its inverse function $f_i^{-1}(\cdot)$ is a one-to-one mapping. Using (3.11), an analytical form of (3.14) is obtained as [4]:

$$\hat{\gamma}_i = c_i - \frac{1}{b_i} \ln\left[\frac{b_i h_{i,i}}{2a_i^{(m)} I_i} - 1 - \sqrt{\left(1 - \frac{b_i h_{i,i}}{2a_i^{(m)} I_i}\right)^2 - 1}\right]. \tag{3.15}$$

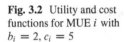

Fig. 3.2 Utility and cost functions for MUE i with $b_i = 2, c_i = 5$

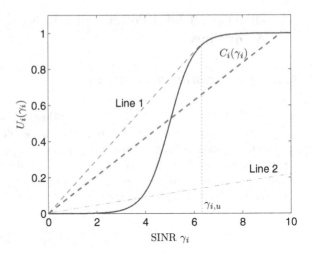

In Fig. 3.2, we plot both utility and cost functions versus SINR γ_i. As the cost function in (3.12) is $C_i(\gamma_i) = \left(a_i^{(m)} I_i / h_{i,i} \right) \gamma_i$, its slope with respect to γ_i is $C_i'(\gamma_i) = a_i^{(m)} I_i / h_{i,i}$. Because $I_i \geq \sigma_i$, $C_i'(\gamma_i)$ has a minimum value of $\underline{z}_i = a_i^{(m)} \sigma_i / h_{i,i}$, which is the slope of line 2 in Fig. 3.2. On the other hand, line 1 that goes through the origin and is tangent to the utility curve $U_i(\gamma_i)$ at $\gamma_{i,u}$ takes the form:

$$U_i(\gamma_{i,u}) = U_i'(\gamma_{i,u}) \gamma_{i,u}. \tag{3.16}$$

For a nonnegative total utility, $C_i(\gamma_i)$ must be below line 1, which means that $C_i'(\gamma_i) \leq \bar{z}_i = U_i'(\gamma_{i,u})$.

Since $\hat{\gamma}_i$ is the solution of $U_i'(\gamma_i) = C_i'(\gamma_i)$, it is the point where $U_i'(\gamma_i)$ and $C_i'(\gamma_i)$ intersect. As shown in Fig. 3.3, any $C_i'(\gamma_i)$ in the interval $[\underline{z}_i, \bar{z}_i]$ will have two intersections with $U_i'(\gamma_i)$. However, we only take the intersection on the right side as the solution $\hat{\gamma}_i$ because this side corresponds to the concave part of $U_i(\gamma_i)$. The intersection on the left side actually gives the *minimum* total utility, and thus is ignored. It is clear from Fig. 3.3 that $\hat{\gamma}_i \geq \gamma_{i,u}$. Therefore, by setting:

$$\gamma_{i,u} = \gamma_i^{\min}, \tag{3.17}$$

we can ensure that any active MUE (i.e., whose transmit power is strictly positive) will attain its minimum SINR target.

Note that $U_i'(\cdot)$ becomes very steep with a sufficiently large b_i. In such cases, Fig. 3.3 shows that the resulting $\hat{\gamma}_i$ of MUE i will be very close to its SINR

Fig. 3.3 Operating region of $\hat{\gamma}_i$ of MUE i with $b_i = 2$, $c_i = 5$

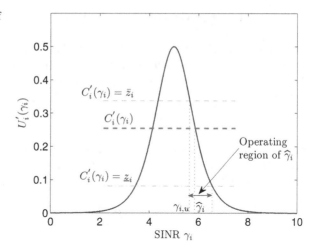

threshold $\gamma_i^{\min}$. Also, if the minimum required SINRs of all MUEs are feasible, we can activate all MUEs by setting $a_i^{(m)}$ sufficiently small. Specifically, given I_i MUE $i \in \mathscr{L}_m$ is active if $C_i'(\gamma_i) < U_i'(\gamma_i^{\min})$, i.e., $a_i^{(m)} < h_{i,i} U_i'(\gamma_i^{\min})/I_i$.

Some manipulations of (3.16) and (3.17) give [4]:

$$c_i = \gamma_i^{\min} - \frac{\ln(b_i \gamma_i^{\min} - 1)}{b_i}. \tag{3.18}$$

By substituting this value of c_i to (3.15), we finally arrive at:

$$\hat{\gamma}_i = \gamma_i^{\min} - \frac{\ln(b_i \gamma_i^{\min} - 1)}{b_i} - \frac{1}{b_i} \ln \left[\frac{b_i h_{i,i}}{2a_i^{(m)} I_i} - 1 - \sqrt{\left(1 - \frac{b_i h_{i,i}}{2a_i^{(m)} I_i}\right)^2 - 1} \right]. \tag{3.19}$$

3.2.2 Dynamic Pricing, Power Adaptation and Admission Control for FUEs

Given MUEs' QoS requirements already supported, the specific choice of utility and cost functions for FUEs allows us to achieve several practical design objectives, through which certain UE satisfaction metrics can be attained. If FUEs also wish to maintain their respective QoS requirements, the operation of these UEs may cause network congestion, hence badly affecting the performance of MUEs. In such cases, FUEs should be penalized by appropriately regulating their operating parameters.

3.2.2.1 Balancing Achieved Throughput and Power Expenditure for FUEs

We choose a utility function that represents the Shannon capacity for FUEs, i.e., $U_i(\gamma_i) = B \ln(1 + \gamma_i)$ where B denotes the system bandwidth, and a linear cost function $C(p_i) = a_i^{(f)} p_i$ with pricing coefficient $a_i^{(f)}$. The net utility of FUE i is:

$$U_{\text{tot},i} = B \ln(1 + \gamma_i) - a_i^{(f)} p_i, \quad \forall i \in \mathscr{L}_f. \tag{3.20}$$

Such choices of functions are especially relevant when FUEs have to tradeoff between achieving the highest possible data rates and expending as little power as necessary. Applying the result in (3.7) to these utility and cost functions gives:

$$\frac{B}{1 + \gamma_i} = \frac{a_i^{(f)} I_i}{h_{i,i}}. \tag{3.21}$$

From (3.21) and upon noting that $U_{\text{tot},i}$ is strictly concave in p_i, the value of $p_i \geq 0$ that globally maximizes $U_{\text{tot},i}$ can be derived as:

$$p_i^* = \left(\frac{B}{a_i^{(f)}} - \frac{I_i}{h_{i,i}} \right)^+. \tag{3.22}$$

We present in Algorithm 3.1 a joint power and admission control scheme for interference management in both macrocell and femtocell networks. This algorithm lends itself to a distributed implementation with only local information required. In every iteration, each UE $i \in \mathscr{L}$ simply needs to estimate (1) its received interference power $I_i[t]$, and (2) its own channel gain $h_{i,i}$ to update its transmit power. When there exists an active MUE i with its "soft" SINR target $\hat{\gamma}_i[t]$ dropping below the prescribed SINR target $\gamma_i^{\min}$, we gradually increase pricing coefficients $a_j^{(f)}$ of all the active FUEs [see Step 7]. It is apparent from (3.22) that such an increase in $a_j^{(f)}$ results in a reduction in the transmit power of FUE j, through which the FUE that creates undue cross-tier interference can be effectively penalized. Notably, this pricing mechanism is realized without acquiring the knowledge of cross-channel gains, unlike the one proposed by [3].

The procedure of updating the pricing coefficients described in Step 7 of Algorithm 3.1 impacts both the convergence speed of the algorithm and the number of active FUEs at equilibrium. A higher initial pricing $a_j^{(f)}$ and/or a larger scaling factor $k_j^{(f)} > 1$ will shorten the convergence time, albeit at the cost of being able to support a fewer number of active FUEs at the equilibrium point. Therefore, careful selections of $k_j^{(f)}$ and $a_j^{(f)}$ to reflect the relative amount of interference that FUE $j \in \mathscr{L}_f$ induces to other UEs may lead to better network performance. In particular, it is sensible to set large values of $k_j^{(f)}$ and $a_j^{(f)}$ for FUE j who creates

---

Algorithm 3.1 Joint power and admission control for macrocell QoS guarantee and femtocell throughput-power tradeoff

---

1: Set $p_i := 0$, $\forall i \in \mathscr{L}$, initialize the set of active FUEs $\mathscr{L}_f^A := \mathscr{L}_f$, and set $t := 1$.
2: Each MUE $i \in \mathscr{L}_m$ measures $h_{i,i}$ and $I_i[t]$, and calculates $\hat{\gamma}_i[t]$ by (3.19).
3: **if** $\hat{\gamma}_i[t] \geq \gamma_i^{\min}$ **then**

4:     MUE $i \in \mathscr{L}_m$ updates its power as $p_i[t+1] := \dfrac{I_i[t]\hat{\gamma}_i[t]}{h_{i,i}}$.

5: **else if** $\hat{\gamma}_i[t] < \gamma_i^{\min}$ and $\left|\mathscr{L}_f^A\right| > 0$ **then**

6:     MUE $i \in \mathscr{L}_m$ updates its power as $p_i[t+1] := \dfrac{I_i[t]\gamma_i^{\min}}{h_{i,i}}$.

7:     Each FUE $j \in \mathscr{L}_f^A$ updates its pricing coefficient: $a_j^{(f)} := k_j^{(f)}a_j^{(f)}$, where $k_j^{(f)} > 1$ are predetermined scaling factors.

8: **end if**
9: Each FUE $j \in \mathscr{L}_f^A$ measures $h_{j,j}$ and $I_j[t]$, calculates $\hat{p}_j$ as:

$$\hat{p}_j := \frac{B}{a_j^{(f)}} - \frac{I_j[t]}{h_{j,j}}.$$

10: FUE $j \in \mathscr{L}_f^A$ updates its power: $p_j[t+1] := \hat{p}_j$.

11: **if** $\dfrac{\hat{p}_j h_{j,j}}{I_j[t]} < \underline{\gamma}_j$ **then**

12:     If $t = nT^{(f)}$ then, with a small probability $\bar{\alpha}$, FUE $j \in \mathscr{A}_f$ sets $p_j[t+1] := 0$ and removes itself from the set of active FUEs: $\mathscr{L}_f^A := \mathscr{L}_f^A \setminus \{j\}$.

13: **end if**
14: Any femtocell BS with no associated active FUE informs the macrocell through a dedicated signaling channel.
15: Set $t := t+1$, go to Step 2 and repeat until convergence.

---

excessive interference. Eventually, these "bad" UEs will at least see their transmit power reduced at equilibrium. In networks with a high load level[1], they can even be removed, through which the built-up network congestion is relieved.

It is desirable not to remove the FUEs prematurely. Therefore, in Step 12 of Algorithm 3.1 if a certain FUE j has its SINR falling below the minimum threshold required for useful communication $\underline{\gamma}_j$, we remove it with probability $\bar{\alpha}$ (where $0 < \bar{\alpha} < 1$) at most once in every $T^{(f)}$ iterations. A small value of $\bar{\alpha}$ will prevent an unnecessary elimination of too many FUEs, albeit at the expense of prolonging the convergence time. The same effect can also be expected for large values of $T^{(f)}$.

---

[1]In this study, the network load is defined to be "low" if $\bar{\rho} = \rho(\mathrm{diag}([\boldsymbol{\gamma}_m^{\min}; \boldsymbol{\gamma}_f^{\min}])\mathbf{H}) < 1$, where $\rho(\cdot)$ denotes the matrix spectral radius, $\mathbf{H} = [h_{i,j}]_{i,j}$ is the channel gain matrix, and $\boldsymbol{\gamma}_m^{\min}$ and $\boldsymbol{\gamma}_f^{\min}$ are the vectors of minimum SINRs required by MUEs and FUEs, respectively. When $\bar{\rho} \geq 1$, not all minimum SINRs can be supported with finite transmit powers [9]. Therefore, the network load level is either "medium" or "high" depending on the specific value of $\bar{\rho}$, and admission control is needed to remove some FUEs. If $\bar{\rho}$ tends to be much larger than 1, the network becomes very congested where it is even difficult to support the minimum SINR $\boldsymbol{\gamma}_m^{\min}$ of the MUEs alone.

Theorem 3.1. *The proposed Algorithm 3.1 converges to an equilibrium solution if* $x f_i^{-1}(x)$ *is an increasing function,* $\forall i \in \mathscr{L}_m$, *and the following condition*

$$(M_{\mathscr{A}} + K_{\mathscr{A}} - 1) \bar{r} < 1 \tag{3.23}$$

holds, where $f_i(\cdot) = U_i'(\cdot)$ *in the concave part of* $U_i(\gamma_i)$; $M_{\mathscr{A}} = \left| \mathscr{L}_m^A \right| \leq M$ *and* $K_{\mathscr{A}} = \left| \mathscr{L}_f^A \right| \leq K$ *denote the cardinality of the sets of active MUEs and FUEs, respectively; and* $\bar{r}$ *is defined as:*

$$\bar{r} = \max_{i \in \mathscr{L}_f, j \in \mathscr{L} \setminus \{i\}} \frac{h_{i,j}}{h_{i,i}} = \max_{i \in \mathscr{L}_f, j \in \mathscr{L} \setminus \{i\}} \frac{h_{i,j}}{G h_{i,i}'}. \tag{3.24}$$

Moreover, for UEs achieving nonzero powers at the equilibrium, it is true that

$$p_i^* = \frac{I_i^*}{h_{i,i}} f_i^{-1} \left(\frac{a_i^{(m)} I_i^*}{h_{i,i}} \right), \quad i \in \mathscr{L}_m^A, \tag{3.25}$$

$$p_i^* = \frac{B}{a_i^{(f)}} - \frac{I_i^*}{h_{i,i}}, \quad i \in \mathscr{L}_f^A, \tag{3.26}$$

where $I_i^* = \sum_{j \neq i} h_{i,j} p_j^* + \sigma_i$. *Further, all active MUEs* $i \in \mathscr{L}_m^A$ *have their SINR* γ_i^* *satisfying* $\gamma_i^* \geq \gamma_i^{\min}$.

Proof. The proof can be found in [2].   □

3.2.2.2 Soft QoS Provisioning for FUEs

In this scenario, we assume that FUE $i \in \mathscr{L}_f$ also requires a minimum SINR $\gamma_i^{\min}$ to maintain the quality of its communication. Note that $\gamma_i^{\min}$ here is different from $\underline{\gamma}_i$ defined in (3.4), with $\gamma_i^{\min}$ typically greater than $\underline{\gamma}_i$ in practice. While a higher SINR at the receiving end of any femto links implies more reliability and better services, this usually requires more transmit power, which in turn leads to a higher cross-interference induced to the macrocell. Such an observation motivates us to consider the following net utility for FUE i [10]:

$$U_{\text{tot},i} = -\left(\gamma_i - \gamma_i^{\min} \right)^2 - a_i^{(f)} p_i, \quad \forall i \in \mathscr{L}_f. \tag{3.27}$$

Although maximizing $U_i(\gamma_i) = -\left(\gamma_i - \gamma_i^{\min} \right)^2$ in (3.27) enforces the SINR γ_i of FUE i to be as close as possible to the SINR target $\gamma_i^{\min}$, the resulting γ_i^* at the equilibrium may actually be less than $\gamma_i^{\min}$. Nevertheless, it is shown in [10] that by allowing a small deviation from the target SINR, a significant reduction in the transmit power (and hence, the resulting interference) can be achieved. Given its lower access priority, this type of soft QoS provisioning is acceptable by FUE i. On the other hand, cost function $C_i(p_i) = a_i^{(f)} p_i$ penalizes the expenditure of transmit power, which potentially creates undue interference to the macrocell as well as other FUEs. Here, $a_i^{(f)}$ is the pricing coefficient of such penalization.

Now, applying the result in (3.7) to these particular utility and cost functions yields:

$$\gamma_i = \gamma_i^{\min} - \frac{a_i^{(f)} I_i}{2h_{i,i}}. \tag{3.28}$$

$U_i(\gamma_i)$ is a concave function in p_i, and so is $U_{\text{tot},i}$, $\forall i \in \mathscr{L}_f$. The power value that globally maximizes $U_{\text{tot},i}$ can thus be computed as:

$$p_i^* = \left(\frac{I_i \gamma_i^{\min}}{h_{i,i}} - \frac{a_i^{(f)} I_i^2}{2h_{i,i}^2} \right)^+. \tag{3.29}$$

Again, by setting the pricing coefficient $a_i^{(f)}$ to be sufficiently large, we can effectively shut off FUE i. Based upon the power update rule in (3.29), a joint power adaptation and admission control algorithm can be developed that is capable of providing soft QoS for FUEs. This algorithm is referred to as Algorithm 3.2 in the sequel. The steps in Algorithm 3.2 are identical to those in Algorithm 3.1, except for Step 9 where $\hat{p}_j$ is instead calculated as:

$$\hat{p}_j = \frac{I_j[t] \gamma_j^{\min}}{h_{j,j}} - \frac{a_j^{(f)} I_j^2[t]}{2h_{j,j}^2}, \quad \forall j \in \mathscr{A}_f. \tag{3.30}$$

Theorem 3.2. *Assuming that $x f_i^{-1}(x)$ is an increasing function, $\forall i \in \mathscr{L}_m$, the proposed Algorithm 3.2 converges to an equilibrium, at which point we have that:*

$$p_i^* = \frac{I_i^*}{h_{i,i}} f_i^{-1} \left(\frac{a_i^{(m)} I_i^*}{h_{i,i}} \right), \ i \in \mathscr{L}_m^A \tag{3.31}$$

$$p_i^* = \frac{I_i^* \gamma_i^{\min}}{h_{i,i}} - \frac{a_i^{(f)} (I_i^*)^2}{2h_{i,i}^2}, \ i \in \mathscr{L}_f^A. \tag{3.32}$$

Moreover, all active MUEs $i \in \mathscr{L}_m$ have their SINR γ_i^* satisfying $\gamma_i^* \geq \gamma_i^{\min}$.

Proof. The proof can be found in [2].   □

3.3 Practical Implementation Issues and Further Extensions

3.3.1 Communication Overhead of Proposed Algorithms

The two algorithms developed in this work only require a limited amount of signaling to be exchanged among the femtocells and macrocell. In either algorithm, the power updates of both MUEs and FUEs can be executed in a completely distributed manner, based on the information available at local links. On one hand, the receiver of each UE i (i.e., either the BS or the UE terminal depending on the uplink or downlink transmission, respectively) can estimate $h'_{i,i}$ by, for instance, exploiting the pilot channel. On the other hand, this receiver can also measure the total received power, and then subtract its own received power to obtain the aggregated interference I_i, i.e., $I_i = \sum_{j \in \mathcal{L}} h_{i,j} p_j - h'_{i,i} p_i$, assuming that noise can be ignored in interference-limited CDMA links. The receiver of UE i then sends both values of $h'_{i,i}$ and I_i to its transmitter for the update of transmit power in each iteration.

In Step 7 of Algorithms 3.1 and 3.2, FUEs are requested to increase their pricing coefficients when certain MUEs perceive network congestion. It is realistic to assume that each MUE may only experience significant interference from the FUEs within its immediate neighborhood. To protect MUEs in the downlink case, it would therefore be sufficient that only the macrocell receivers with low SINRs request their neighboring FUEs to increase their pricing coefficients. In the case of open access, a hand-off procedure should be established between UEs and macrocell/femtocell BSs, with a control channel dedicated for this purpose. Here, the "warning" message that asks for an increase in the FUEs' prices can be incorporated into the hand-off message when undue cross-tier interference is sensed by the victim MUEs. The other type of communication overhead includes the notification made by the femtocell BS that serves no FUEs to the macrocell in Step 14 of Algorithms 3.1 and 3.2. This may take the form of a simple flag message, to be sent over the available wired backhaul network or be broadcast wirelessly.

3.3.2 Improving Efficiency of Equilibrium Solutions

The equilibrium solutions achieved by the developed algorithms correspond to the Nash equilibria of the underlying non-cooperative games [11]. In such states, no UE has any incentive to unilaterally change its transmit power level. However, Nash equilibrium in general does not guarantee to be either globally efficient or optimal. We discuss here a mechanism to improve the efficiency of this equilibrium, particularly when the system is in lightly-loaded condition. Specifically, we attempt to make the SINRs of active MUEs greater than their required SINRs. In wireless environments, this result implies more service reliability and more robustness against fading for MUEs.

From (3.13), the following relationship at the equilibrium is obtained for an active MUE i:

$$f_i(\gamma_i^*) = \frac{a_i^{(m)} I_i^*}{h_{i,i}}, \tag{3.33}$$

where a typical shape of function $f_i(\cdot)$ has already been illustrated in Fig. 3.3. It is observed that a higher SINR γ_i^* can be realized for a given $a_i^{(m)} I_i^*/h_{i,i}$ if $f_i(\cdot)$ becomes flatter. This corresponds to choosing smaller values of b_i, where recall that b_i is the parameter controlling the steepness of $U_i(\gamma_i)$. Ultimately, it is possible that the SINRs of MUEs are enhanced by reducing b_i whenever possible. As we also need large b_i's for $x f_i^{-1}(x)$ to be increasing, the values of b_i should be updated less frequently compared with the update of power itself.

Towards this end, the following procedure can be employed to improve the attained SINRs of active MUEs: we choose in advance a particular interval T_b to periodically update $b_i, \forall i \in \mathscr{L}_m$. At the beginning of each interval T_b, MUE $i \in \mathscr{L}_m$ multiplies b_i by a factor $k_b < 1$ if its servicing macrocell BS has not been informed about any empty femtocell during the previous interval. The latter situation happens if the network load is low, which also means that most of the FUEs converge to their desired equilibrium without being removed.

3.4 Illustrative Results

The network setting and UE placement in our numerical examples are illustrated in Fig. 3.4, where MUEs and FUEs are randomly deployed inside circles of radii of 500 m and 100 m, respectively. For the ease of reference, the simulation parameters are summarized in Table 3.1. Downlink transmission is considered in all the simulations. We also assume that the number of FUEs serviced by any femtocell BS is identical. The specific numbers of MUEs and FUEs generated in each example are displayed under the plots. The results presented in each figure correspond to one particular network realization, chosen with the intention to demonstrate certain features of the developed algorithms. The channel gain from the transmitter of UE j to the receiver of UE i is calculated as $d_{i,j}^{-\beta}$, where $d_{i,j}$ is their geographical distance and β the pathloss exponent. The same initial pricing coefficient $a_i^{(f)} = a^{(f)}$ and scaling parameter $k_i^{(f)} = k^{(f)}$ are used for all FUEs. Their values, together with SINR targets $\gamma_i^{\min}$, can be found underneath every plot. In each figure, a single curve corresponds to one specific UE.

In Fig. 3.5a, b, we show the evolutions of powers and SINRs under Algorithm 3.1. As can be seen, Algorithm 3.1 converges to an equilibrium with the target SINRs being attained for all the MUEs. It is also clear from these figures that the convergence time of Algorithm 3.1 is relatively short, slightly larger than 10 iterations in this case. On the other hand, Fig. 3.5c illustrates

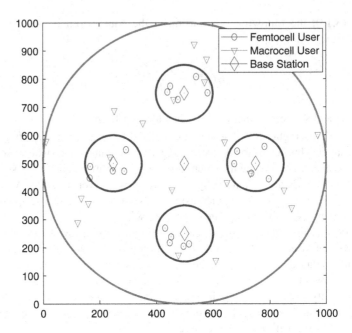

Fig. 3.4 Network topology and user placement in the numerical examples

Table 3.1 Simulation parameters

Parameter	Value
Path-loss exponent, β	3
Processing gain, G	100
Noise power, $\sigma_i = \sigma$ (in W) $\forall i \in \mathcal{L}$	10^{-10}
System bandwidth, B (in Hz)	10^6
γ_i	2
$a_i^{(m)} = a_i, \forall i \in \mathcal{L}_m$	1
b_i	1
Removal probability, $\bar{\alpha}$	0.1
$T^{(f)}$	10
T_b	20
k_b	0.5

the operation of Algorithm 3.1 when the network becomes congested with $\rho\big(\mathrm{diag}([\gamma_m^{\min}; \gamma_f^{\min}])\mathbf{H}\big) = 1.1$. This algorithm initially converges to an equilibrium in which the SINR requirement of one FUE cannot be satisfied, i.e., its final SINR drops below the threshold $\gamma = 2$. Then, the admission control mechanism integrated in Algorithm 3.1 is engaged to effectively remove this UE, resulting in a noticeable growth in SINRs of several other FUEs [see iteration 20 and beyond]. It is also evident here that the removal of FUEs does not affect the transmit powers and SINRs of MUEs. This result verifies the efficiency and robustness of Algorithm 3.1 in protecting the macrocell performance.

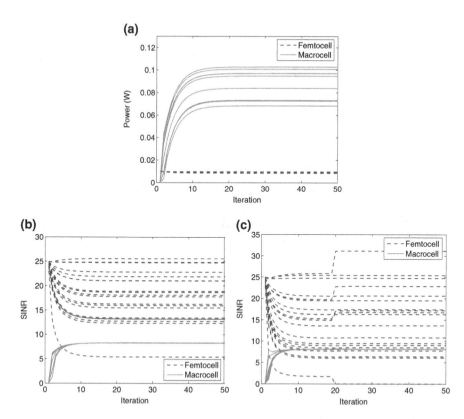

Fig. 3.5 Performance of Algorithm 3.1 with $M = 10$, $K = 20$, $\gamma_i^{min} = 8$ $(i \in \mathcal{L}_m)$, $a^{(f)} = 10^9$ and $k^{(f)} = 1.1$. (**a**) Power evolution. (**b**) SINR evolution. (**c**) SINR evolution with FUE removal

In Fig. 3.6a–c, we display the evolutions of SINRs for all UEs under Algorithm 3.2 when the network load level is low, medium, and high, respectively. In all the scenarios, it is confirmed that Algorithm 3.2 actually converges with the SINR requirements of all MUEs being met at the equilibrium. When the network becomes more congested, the convergence speed appears to be slower. Specifically, Fig. 3.6a shows that when the network load is low (i.e., $\rho\big(\mathrm{diag}([\boldsymbol{\gamma}_m^{min}; \boldsymbol{\gamma}_f^{min}])\mathbf{H}\big) = 0.9$), the achieved SINRs of FUEs are slightly below their corresponding requirements while the performance of all MUEs is well protected. This is a desirable feature as a "soft" QoS for the lower-tier FUEs can only be supported to the extent that network load allows. When network congestion starts building up, Algorithm 3.2 smoothly reduces the SINRs of FUEs so that MUEs can eventually reach their desired SINR targets. This feature can best be observed in Fig. 3.6b, where we set $\rho\big(\mathrm{diag}([\boldsymbol{\gamma}_m^{min}; \boldsymbol{\gamma}_f^{min}])\mathbf{H}\big) = 1.1$. Finally, when the network gets so congested (i.e., $\rho\big(\mathrm{diag}([\boldsymbol{\gamma}_m^{min}; \boldsymbol{\gamma}_f^{min}])\mathbf{H}\big) = 1.18$) that the SINRs of certain FUEs fall below the minimum required threshold $\underline{\gamma} = 2$, admission control is executed to remove such UEs. This operation is depicted in Fig. 3.6c, where the FUE that achieves the smallest SINR value is eliminated from the network.

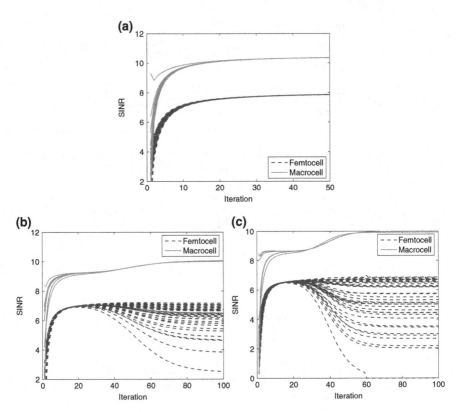

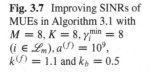

Fig. 3.6 Performance of Algorithm 3.2 with $M = 10, K = 40, \gamma_i^{\min} = 10 \ (i \in \mathscr{L}_m), \gamma_i^{\min} = 8 \ (i \in \mathscr{L}_f), a^{(f)} = 10^4$ and $k^{(f)} = 1.5$. (**a**) Low load. (**b**) Medium load. (**c**) High load

Fig. 3.7 Improving SINRs of MUEs in Algorithm 3.1 with $M = 8, K = 8, \gamma_i^{\min} = 8$ $(i \in \mathscr{L}_m), a^{(f)} = 10^9,$ $k^{(f)} = 1.1$ and $k_b = 0.5$

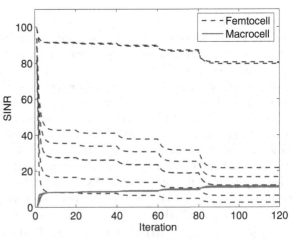

Figure 3.7 illustrates how the technique presented in Sect. 3.3.2 can help improve the achieved SINRs of MUEs in Algorithm 3.1. Recall that such a mechanism, which involves scaling down the values of b_i over time, may only be activated when the network load level is low. To obtain the results presented in this figure, we have decreased all b_i's by a factor $k_b = 0.5$ once in every $T_b = 20$ iterations. These updates are carried on until one FUE settles its SINR below the specified SINR threshold $\underline{\gamma} = 2$. Furthermore, T_b is set to be sufficiently large so that the algorithm converges to a new equilibrium. As shown in Fig. 3.7, by scaling down b_i, we can enhance the attained SINRs of all MUEs at the cost of degrading the SINRs of FUEs.

References

1. D. T. Ngo, L. B. Le, T. Le-Ngoc, E. Hossain, and D. I. Kim, "Distributed interference management in femtocell networks," in *Proc. IEEE Veh. Tech. Conf. (VTC-Fall)*, San Franciso, CA, Sep. 2011, pp. 1–5.
2. D. T. Ngo, L. B. Le, T. Le-Ngoc, E. Hossain, and D. I. Kim, "Distributed interference management in two-tier CDMA femtocell networks," *IEEE Trans. Wireless Commun.*, vol. 11, no. 3, pp. 979–989, Mar. 2012.
3. V. Chandrasekhar, J. G. Andrews, T. Muharemovic, and Z. Shen, "Power control in two-tier femtocell networks," *IEEE Trans. Wireless Commun.*, vol. 8, no. 8, pp. 4316–4328, Aug. 2009.
4. M. Xiao, N. B. Shroff, and E. K. P. Chong, "A utility-based power control scheme in wireless cellular systems," *IEEE/ACM Trans. Netw.*, vol. 11, no. 2, pp. 210–221, Apr. 2003.
5. G. J. Foschini and Z. Miljanic, "A simple distributed autonomous power control algorithm and its convergence," *IEEE Trans. Veh. Technol.*, vol. 42, no. 4, pp. 641–646, Nov. 1993.
6. J. Zander, "Distributed cochannel interference control in cellular radio systems," *IEEE Trans. Veh. Technol.*, vol. 41, no. 3, pp. 305–311, Aug. 1992.
7. M. Andersin, Z. Rosberg, and J. Zander, "Gradual removals in cellular PCS with constrained power control and noise," *Wireless Netw.*, vol. 2, pp. 27–43, 1996.
8. N. Bambos, S. C. Chen, and G. J. Pottie, "Channel access algorithms with active link protection for wireless communication networks with power control," *IEEE/ACM Trans. Netw.*, vol. 8, no. 5, pp. 583–597, Oct. 2000.
9. J. Zander, "Performance of optimum transmitter power control in cellular radio systems," *IEEE Trans. Veh. Technol.*, vol. 41, no. 1, pp. 57–62, Feb. 1992.
10. S. Koskie and Z. Gajic, "A Nash game algorithm for SIR-based power control in 3G wireless CDMA networks," *IEEE/ACM Trans. Netw.*, vol. 13, no. 5, pp. 1017–1026, Oct. 2005.
11. D. Fudenberg and J. Tirole, *Game Theory*. MIT Press, 1991.

Chapter 4
Distributed Pareto-Optimal Power Control for Utility Maximization in Heterogeneous CDMA Small-Cell Networks

Chapter 3 has presented joint power and admission control schemes for distributed interference management in two-tier networks [1, 2]. It has been shown that the underlying games settle at some NE, at which point no UE has any incentive to unilaterally change its power level. Although an NE gives a steady operating point, it is generally not guaranteed to be Pareto-efficient. To improve the efficiency of NE solutions, a number of pricing schemes are adopted in [3–8]. Under the proposed load-spillage framework, [9] devises distributed Pareto-optimal solutions to jointly optimize SINRs and powers. Note that all of these studies assume homogeneous networks, where there exist no differentiated classes of users with distinct access priorities and diverse QoS requirements.

On the contrary, it is imperative to protect the ongoing operation of the preferential MUEs at all times in a two-tier heterogeneous network. This critical requirement poses a major challenge that hinders the successful application of any available solutions. Specifically, the choices of target SINRs available to the lower-tier FUEs in this case are much more limited, further complicating the Pareto-optimal boundary of the feasible SINR region. As a direct consequence, locating a particular SINR point on such a boundary to optimize certain system-wide design criteria is by no means a trivial task.

Directly targeting such a central issue, this chapter attempts to develop interference management solutions wherein (i) all UEs attain their respective SINRs that are always optimal in the Pareto sense, and (ii) macrocell and femtocell networks have their utilities globally maximized [10–12]. To handle the above-mentioned QoS requirements of the prioritized MUEs, the Joint Utility Maximization with macrocell Quality-of-Service guarantee (JUM-QoS) algorithm is first proposed that maximizes the total utility of both macrocell and femtocell networks. In particular, the minimum SINRs prescribed by MUEs are effectively enforced with the use of a log-barrier penalty function. After this key step, the Pareto-optimal boundary of the strongly-coupled feasible SINR region is characterized, and the load-spillage framework [9] is specifically adapted to find the

D.T. Ngo and T. Le-Ngoc, *Architectures of Small-Cell Networks and Interference Management*, SpringerBriefs in Computer Science, DOI 10.1007/978-3-319-04822-2__4, © The Author(s) 2014

SINR that approximately maximizes the sum utility. Finally, the global optimum of the original problem is attained by properly tuning the penalty parameter in the proposed penalty approach.

In the specific case where MUEs only need to be assured with some predefined minimum SINRs, the Femtocell Utility Maximization with Macrocell SINR Balancing (FUM-MSB) algorithm is devised. Upon observing the structure of the objective function and the monotonicity of SINR, the Pareto-optimal SINR boundary is confined to a much smaller space. Only then the load-spillage parametrization is applied to FUEs, whereas the loads of all MUEs are updated according to a newly developed iterative procedure. Still operating on the Pareto-optimal SINR frontier and retaining the global optimality, this algorithm outperforms the general counterpart JUM-QoS in several important aspects, including scalability, computational complexity, convergence behavior, and stability around the optimum.

It is noteworthy that the proposed JUM-QoS algorithm can also control the access priority of both the macrocell and femtocells by granting a proper weight to each class of users. In the two developed algorithms, the adopted α-fair utility function can always be regulated to give different degrees of fairness in allocating radio resources to individual UEs. Moreover, the devised schemes can be locally executed, incurring little signaling and information exchange. This feature is particularly attractive in view of practical implementation under the limited backhaul network capacity available for femtocells.

4.1   System Model and Problem Formulation

Consider the *uplink* of a two-tier wireless network, in which M MUEs establish communication with its servicing macrocell BS while K FUEs also transmit to their respective femtocell BSs. Although we assume that all MUEs and FUEs share the same radio frequency bands by CDMA, the results obtained in this chapter are applicable to single-carrier wireless systems in general. Also assume that the association of a certain FUE with its own femtocell BS is fixed during the runtime of the underlying power control. Without loss of generality, denote the set of MUEs and FUEs by $\mathscr{L}_m := \{1, \ldots, M\}$ and $\mathscr{L}_f := \{M + 1, \ldots, M + K\}$, respectively. The set of all UEs is then $\mathscr{L} := \mathscr{L}_m \cup \mathscr{L}_f$, whose cardinality is $|\mathscr{L}| = M + K$. An example of the system under investigation is illustrated in Fig. 4.1. It is assumed here that time scale of network topology changes is very small compared to that of power adaptation. In addition, data transmission time scale is far shorter than that of the underlying optimization process, which allows any short-term statistical variations to be averaged out (see, e.g., [13]).

Denote by θ_i the serving BS of UE $i \in \mathscr{L}$ (which is either an MUE or an FUE). For brevity, the path between UE i and its servicing BS θ_i shall be referred to as link i. Also, let $\bar{h}_{k,j}$ be the absolute channel gain from UE j to BS k, and define its corresponding normalization as $h_{k,j} := \bar{h}_{k,j} / \bar{h}_{\theta_j,j}$. To represent the normalized

Fig. 4.1 Example of a
heterogeneous CDMA-based
wireless network

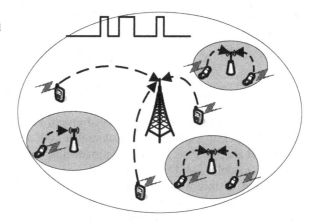

channel gain from UE j to the serving BS θ_i of UE i, we define an $(M + K) \times (M + K)$ channel matrix $\mathbf{H}$ with its (i, j)-th entry being:

$$H_{i,j} := \begin{cases} 0, & \text{if } i = j, \\ 1, & \text{if } \theta_i = \theta_j,\, i \neq j, \\ h_{\theta_i,j}, & \text{if } \theta_i \neq \theta_j. \end{cases} \tag{4.1}$$

Suppose that UE j transmits to its serving BS θ_j, and let $p^{(j)}$ be the received power at θ_j. Since $\bar{h}_{\theta_j,j}$ is the channel gain from j to θ_j, it is clear that j must have transmitted at a power level $p^{(j)}/\bar{h}_{\theta_j,j}$. At any BS k, this signal appears with a power $\bar{h}_{k,j}\left(p^{(j)}/\bar{h}_{\theta_j,j}\right) = h_{k,j}\, p^{(j)}$. The total interference plus noise at BS θ_i that serves UE $i \in \mathscr{L}$ on link i can be expressed as:

$$q^{(i)} := \sum_{j=1}^{M+K} H_{i,j}\, p^{(j)} + \sigma^{(i)}, \tag{4.2}$$

where $\sigma^{(i)}$ is the noise power at the receiving end of link i. Throughout this chapter, we make a reasonable assumption that $\boldsymbol{\sigma} = \left[\sigma^{(1)}, \ldots, \sigma^{(M+K)}\right]^T \neq \mathbf{0}$. In a vector-matrix form, (4.2) can also be written as:

$$\mathbf{q} = \mathbf{H}\mathbf{p} + \boldsymbol{\sigma}. \tag{4.3}$$

Let $\bar{\gamma}^{(i)} := G p^{(i)}/q^{(i)}$ denote the SINR at link $i \in \mathscr{L}$, where G is the system processing gain. For notational convenience, we define the normalized SINR at link i as $\gamma^{(i)} := \bar{\gamma}^{(i)}/G$. It is then easy to see that:

$$\mathbf{p} = \mathbf{D}(\boldsymbol{\gamma})\mathbf{q}, \tag{4.4}$$

where $\mathbf{D}(\gamma) := \mathrm{diag}\left(\gamma^{(1)}, \ldots, \gamma^{(M+K)}\right)$. By substituting (4.4) to (4.3) and after some simple algebra, we yield [9]:

$$\mathbf{q} = \mathbf{HD}(\gamma)\mathbf{q} + \sigma, \tag{4.5}$$

$$\mathbf{p} = \mathbf{D}(\gamma)\mathbf{Hp} + \mathbf{D}(\gamma)\sigma. \tag{4.6}$$

Since we do not consider totally isolated groups of links that are not interacting with each other, it is practical to assume that both non-negative matrices $\mathbf{HD}(\gamma)$ and $\mathbf{D}(\gamma)\mathbf{H}$ are *primitive*, i.e., they are irreducible and have only one eigenvalue of maximum modulus [14, Definition 8.5.0].

The goal of this work is to devise jointly optimal power allocation $\mathbf{p}$ and SINR assignment γ solutions for the two types of users (i.e., MUEs and FUEs) with different service priorities and QoS requirements. The prioritized MUEs with higher access rights demand that their ongoing services be, at least, unaffected regardless of any femtocell deployment. Therefore, a set of minimum SINRs $\gamma^{\min} = [\gamma^{(1)\min}, \ldots, \gamma^{(M)\min}]^T$ prescribed by the MUEs must always be maintained:

$$\gamma^{(i)} \geq \gamma^{(i)\min}, \quad \forall i \in \mathscr{L}_m, \tag{4.7}$$

where $\gamma^{(i)\min}$ is the normalized target SINR corresponding to the actual SINR $\bar{\gamma}^{(i)\min} = G\gamma^{(i)\min}$ required by MUE i. Note that a general QoS $\gamma^{\min}$ can be translated to different specific requirements. For instance, a higher value of $\gamma^{(i)\min}$ means that a higher throughput, a lower BER, and a shorter time delay are guaranteed for MUE i.

Our design objective is to maximize the sum utility of all UEs. Typically an increasing function, utility $U_i(\gamma^{(i)})$ represents the value that UE $i \in \mathscr{L}$, who is assigned with SINR $\gamma^{(i)}$, contributes to the overall network. The higher the SINR, the greater the contribution. Depending on the type of utility functions, fairness, an important system-wide objective, can also be achieved. Proportional fairness and max-min fairness are among the most common metrics used in practice to characterize how competing users share system resources. The α-fair function proposed by [15] provides a useful means to enforce these two types of fairness, in that it generalizes proportional fairness and includes arbitrarily close approximations of max-min fairness. Specifically, we consider the following utility for UE $i \in \mathscr{L}$:

$$U_i(\gamma^{(i)}) := \begin{cases} \log(\gamma^{(i)}), & \text{if } \alpha = 1 \\ (1-\alpha)^{-1}(\gamma^{(i)})^{1-\alpha}, & \text{if } \alpha \geq 0 \text{ and } \alpha \neq 1. \end{cases} \tag{4.8}$$

Here, $\alpha = 1$ corresponds to proportional fairness whereas $\alpha \to \infty$ gives max-min fairness.

Let $\rho(\mathbf{X})$ denote the spectral radius of the matrix $\mathbf{X}$, i.e., the maximum modulus eigenvalue of $\mathbf{X}$. Given the channel matrix $\mathbf{H}$, the specific value of $\rho(\mathbf{HD}(\gamma))$ indicates whether a certain SINR γ is supportable. In particular, it is required that $\rho(\mathbf{HD}(\gamma)) < 1$ for the existence of a feasible power vector $\mathbf{p} \succ \mathbf{0}$ [16]. In the limit that $\rho\left(\mathbf{HD}(\boldsymbol{\gamma})\right) = 1$, an infinite amount of transmit power is needed to attain $\boldsymbol{\gamma}$. For $\rho\left(\mathbf{HD}(\boldsymbol{\gamma})\right) > 1$, the network can be regarded so congested that only removing certain UEs and/or lowering the SINR targets can help relieve such congestion. Considering a practical non-congested network with attainable target SINRs, we insist that $\rho\left(\mathbf{HD}(\boldsymbol{\gamma})\right) \leq \bar{\rho}$ where $0 \leq \bar{\rho} < 1$, for the existence of a feasible solution with $0 < p^{(i)} < \infty$, $\forall i \in \mathscr{L}$.

Given $\bar{\rho} \in [0, 1)$, we are interested in the following problem:

$$\max_{\boldsymbol{\gamma} \in \mathbb{R}_+^{M+K}, \, \mathbf{p} \in \mathbb{R}_+^{M+K}} \quad w_m \sum_{i \in \mathscr{L}_m} U_i(\gamma^{(i)}) + w_f \sum_{i \in \mathscr{L}_f} U_i(\gamma^{(i)})$$

$$\text{s.t.} \quad \rho\left(\mathbf{HD}(\boldsymbol{\gamma})\right) \leq \bar{\rho}, \tag{4.9}$$

$$\gamma^{(i)} \geq \gamma^{(i)\,\text{min}}, \, \forall i \in \mathscr{L}_m$$

where $w_m \geq 0$ and $w_f \geq 0$ designate the importance toward the macrocell and femtocell network, respectively. Note that a larger value of $\bar{\rho}$ corresponds to a larger feasible set, and in turn a potentially higher utility. Therefore, it is desirable to choose $\bar{\rho}$ to be as close to 1 as possible while ensuring that $\boldsymbol{\gamma}$ be supportable.

Problem (4.9) is not convex because the set $\{\boldsymbol{\gamma} \in \mathbb{R}_+^{M+K} \mid \rho\left(\mathbf{HD}(\boldsymbol{\gamma})\right) \leq \bar{\rho}\}$ is not convex. However, if we let $\boldsymbol{\Gamma} := \log \boldsymbol{\gamma}$ then its equivalence $\{\boldsymbol{\Gamma} \in \mathbb{R}^{M+K} \mid \rho\left(\mathbf{HD}(e^{\boldsymbol{\Gamma}})\right) \leq \bar{\rho}\}$ is actually a convex set [17, Theorem 1]. Through such a change of variable and upon denoting $\Gamma^{(i)\,\text{min}} := \log(\gamma^{(i)\,\text{min}})$, the following equivalent problem of (4.9) is considered instead:

$$\max_{\boldsymbol{\Gamma} \in \mathbb{R}^{M+K}, \, \mathbf{p} \in \mathbb{R}_+^{M+K}} \quad w_m \sum_{i \in \mathscr{L}_m} U_i(\Gamma^{(i)}) + w_f \sum_{i \in \mathscr{L}_f} U_i(\Gamma^{(i)})$$

$$\text{s.t.} \quad \rho\left(\mathbf{HD}(e^{\boldsymbol{\Gamma}})\right) \leq \bar{\rho}, \tag{4.10}$$

$$\Gamma^{(i)} \geq \Gamma^{(i)\,\text{min}}, \, \forall i \in \mathscr{L}_m.$$

In this case, the utility function becomes:

$$U_i(\Gamma^{(i)}) = \begin{cases} \Gamma^{(i)}, & \text{if } \alpha = 1 \\ (1-\alpha)^{-1} e^{(1-\alpha)\Gamma^{(i)}}, & \text{if } \alpha \geq 0 \text{ and } \alpha \neq 1, \end{cases} \tag{4.11}$$

which is increasing, twice-differentiable and concave with respect to $\Gamma^{(i)}$. Problem (4.10) is a convex optimization program. However, due to the complicated coupling in the feasible region, centralized algorithms are typically needed to resolve (4.10). Given the nature of two-tier networks where central coordination and processing is usually inaccessible, we aim at developing optimal solutions that can be distributively implemented by individual UEs.

4.2   Distributed Power Control for Joint Utility Maximization with Macrocell QoS Protection

4.2.1   Pareto-Optimal SINR Boundary and Approximate Solution via Log-Barrier Penalty Method

We approach problem (4.10) by finding the Pareto-optimal boundary[1] of the feasible SINR region, followed by adapting power to achieve such SINRs. It is therefore imperative to characterize that boundary, through which the coupling can be revealed, allowing for the realization of any distributed mechanisms.

Proposition 4.1. *The Pareto-optimal SINRs for problem (4.10) lie on the following boundary:*

$$\partial \mathcal{G}_{\bar{\rho}} := \left\{ \boldsymbol{\Gamma} \in \mathbb{R}^{M+K} \text{ s.t. } \rho\left(\mathbf{HD}(e^{\boldsymbol{\Gamma}}) \right) = \bar{\rho} \text{ and } \Gamma^{(i)} \geq \Gamma^{(i)\,\text{min}}, \forall i \in \mathcal{L}_m \right\}. \tag{4.12}$$

Proof. The proof can be found in [12]. □

Proposition 4.1 indicates that the search space for Pareto-optimal SINR assignments of (4.10) is $\mathbb{R}^{M+K}$, confined within the surface $\partial \mathcal{G}_{\bar{\rho}}$ specified by $\bar{\rho}$ and $\Gamma^{(i)\,\text{min}}, \forall i \in \mathcal{L}_m$. Consider a simple 3-UE network, Fig. 4.2 illustrates the Pareto-optimal SINR boundaries in both homogeneous and heterogeneous scenarios.

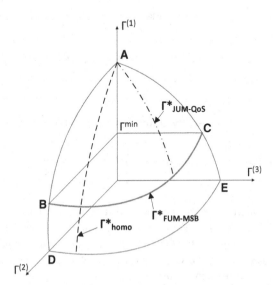

Fig. 4.2 Pareto-optimal SINR boundary of a network consisting of 1 MUE (i.e., UE 1 with $\Gamma^{(1)} \geq \Gamma^{\text{min}}$) and 2 FUEs (i.e., UEs 2 and 3)

---

[1]A feasible SINR $\boldsymbol{\Gamma}$ is called *Pareto-optimal* if it is impossible to increase the SINR of any one link without simultaneously reducing the SINR of some other link.

As can be seen from this example, the optimal SINRs in the heterogeneous case are limited to the smaller region ABC of the surface ADE that contains all possible Pareto-optimal SINRs in the homogeneous case. Locating a particular SINR point on the boundary ABC (i.e., $\partial \mathscr{G}_{\bar{\rho}}$) that maximizes the objective of (4.10) is not trivial. The solution in [9], originally developed for homogeneous networks, is not directly applicable here. It might happen that the SINR point given by [9] lies strictly within the surface BCDE, i.e., outside $\partial \mathscr{G}_{\bar{\rho}}$. In this situation, the successful application of the load-spillage approach relies heavily upon how the critical QoS requirements $\Gamma^{(i)} \geq \Gamma^{(i)\,\mathrm{min}}$, $\forall i \in \mathscr{L}_m$ are managed.

We propose to represent these QoS constraints by the following function:

$$I_-(\Gamma^{(i)}) := \begin{cases} 0, & \text{if } \Gamma^{(i)} \geq \Gamma^{(i)\,\mathrm{min}} \\ \infty, & \text{otherwise,} \end{cases} \tag{4.13}$$

for all $i \in \mathscr{L}_m$. Problem (4.10) thus becomes:

$$\max_{\Gamma \in \mathbb{R}^{M+K},\, \mathbf{p} \in \mathbb{R}_+^{M+K}} \quad w_m \sum_{i \in \mathscr{L}_m} U_i(\Gamma^{(i)}) + w_f \sum_{i \in \mathscr{L}_f} U_i(\Gamma^{(i)}) - \sum_{i \in \mathscr{L}_m} I_-(\Gamma^{(i)}) \tag{4.14}$$

$$\text{s.t.} \quad \rho\left(\mathbf{HD}(e^{\Gamma})\right) \leq \bar{\rho}.$$

Because of (4.13), the objective function in (4.14) is however not differentiable. Therefore, we approximate $I_-(\Gamma^{(i)})$ by:

$$\hat{I}_-(\Gamma^{(i)}) := \begin{cases} -\frac{1}{a} \log\left(\Gamma^{(i)} - \Gamma^{(i)\,\mathrm{min}}\right), & \text{if } \Gamma^{(i)} > \Gamma^{(i)\,\mathrm{min}} \\ \infty, & \text{otherwise,} \end{cases} \tag{4.15}$$

where $a > 0$ is the "penalty factor" used to control the accuracy of the above approximation. Specifically, the approximation becomes more accurate as a increases. $\hat{I}_-(\Gamma^{(i)})$ is convex, non-increasing and differentiable, which also implies the concavity of the objective function in (4.14). Let $\bar{\Phi}(\Gamma) := -\sum_{i \in \mathscr{L}_m} \log(\Gamma^{(i)} - \Gamma^{(i)\,\mathrm{min}})$, whose domain is $\{\Gamma \in \mathbb{R}^{M+K} \mid \Gamma^{(i)} > \Gamma^{(i)\,\mathrm{min}}, \forall i \in \mathscr{L}_m\}$. The following problem, which approximates (4.14), can then be considered:

$$\max_{\Gamma \in \mathbb{R}^{M+K},\, \mathbf{p} \in \mathbb{R}_+^{M+K}} \quad a\left[w_m \sum_{i \in \mathscr{L}_m} U_i(\Gamma^{(i)}) + w_f \sum_{i \in \mathscr{L}_f} U_i(\Gamma^{(i)}) \right] - \bar{\Phi}(\Gamma)$$

$$\text{s.t.} \quad \rho\left(\mathbf{HD}(e^{\Gamma})\right) \leq \bar{\rho}. \tag{4.16}$$

With the proposed penalty function $\bar{\Phi}(\Gamma)$, macrocell SINR constraints are effectively eliminated from the constraint set. This leaves the Pareto-optimal SINR surface be simply $\{\Gamma \in \mathbb{R}^{M+K} \mid \rho\left(\mathbf{HD}(e^{\Gamma})\right) = \bar{\rho}\}$. To distributively realize all points on that surface, we can now make use of the load-spillage framework [9] and parameterize Γ through a new variable $\mathbf{s} \succ \mathbf{0}$ such that $\mathbf{s}^T \mathbf{HD}(e^{\Gamma}) = \bar{\rho}\mathbf{s}^T$. If we let $\mathbf{v} := \mathbf{H}^T \mathbf{s}$, then the resulting SINR

$$\Gamma^{(i)} = \log\left(\bar{\rho}\,s^{(i)}/v^{(i)}\right), \quad \forall i \in \mathscr{L} \tag{4.17}$$

is Pareto-optimal. Since $\mathbf{v}^T \mathbf{D}(e^{\boldsymbol{\Gamma}})\mathbf{H} = \bar{\rho}\mathbf{v}^T$, it is also clear that $\mathbf{v}$ is the left eigenvector of $\mathbf{D}(e^{\boldsymbol{\Gamma}})\mathbf{H}$ with an associated eigenvalue $\bar{\rho}$. Here, $\mathbf{s}$ can be interpreted as the "load" on the network to support an SINR $\boldsymbol{\Gamma}$, whereas $\mathbf{v}$ the "spillage," i.e., the potential interference due to $\boldsymbol{\Gamma}$. More specifically, $v^{(i)} = \sum_j H_{j,i} s^{(j)}$ represents the effect of interference induced by link i to all other links, weighted by the load $s^{(j)}$ of each link. Responsible for the spillage $v^{(i)}$ to achieve a given SINR $\Gamma^{(i)}$, link i loads the network with $s^{(i)} = v^{(i)} e^{\Gamma^{(i)}}/\bar{\rho}$, i.e., it is less tolerant by a factor of $s^{(i)}$ to the interference incurred by other links [9].

By fixing $\bar{\rho} \in [0, 1)$ and upon applying the above parametrization, (4.14) can be solved via an equivalent problem in the variable $\mathbf{s}$. The resolution of this new problem can be accomplished by updating $\mathbf{s}$, taking into account the penalty factor a and $\boldsymbol{\Gamma}^{\min}$, as:

$$s^{(i)}[t+1] := s^{(i)}[t] + \delta\,\Delta s^{(i)}[t+1], \quad \forall i \in \mathscr{L} \tag{4.18}$$

where $\delta > 0$ is a scalar step size, and the search directions for each type of UEs are determined according to:

$$\Delta s^{(i)}[t+1] := \frac{w_m U_i'(\Gamma^{(i)})}{\bar{\rho} q^{(i)}} + \frac{1}{a\bar{\rho}q^{(i)}(\Gamma^{(i)} - \Gamma^{(i)\min})} - s^{(i)}[t], \quad i \in \mathscr{L}_m \tag{4.19}$$

$$\Delta s^{(i)}[t+1] := \frac{w_f U_i'(\Gamma^{(i)})}{\bar{\rho} q^{(i)}} - s^{(i)}[t], \quad i \in \mathscr{L}_f. \tag{4.20}$$

By similar arguments used in [9, Theorems 3 & 4], it can be shown that:

$$\nabla \bar{U}^T \Delta \mathbf{s} = (\partial \bar{U}/\partial \boldsymbol{\Gamma})^T (\partial \boldsymbol{\Gamma}/\partial \mathbf{s}) \Delta \mathbf{s} > 0, \quad \forall \mathbf{s} \succ \mathbf{0}, \tag{4.21}$$

where $\bar{U}(\mathbf{s})$ denotes the objective function of (4.16). This means that (4.19) and (4.20) actually represents an ascent search direction of $\bar{U}(\mathbf{s})$.

In the log-barrier penalty method, it is important to ensure that the conditions $\Gamma^{(i)} > \Gamma^{(i)\min}$ be always satisfied for all $i \in \mathscr{L}_m$ after every update step. Otherwise, the resulting $\boldsymbol{\Gamma}$ would lie outside the domain of $\bar{\Phi}(\boldsymbol{\Gamma})$, making the objective function of (4.16) unbounded from below. To this end, as long as $\Gamma^{(i)} \leq \Gamma^{(i)\min}$ for any $i \in \mathscr{L}_m$, we propose to scale the step size δ in (4.18) as $\delta := b\delta$ where $0 < b < 1$.

Proposition 4.2. *As $\bar{\rho} \to 1$, the update of $\mathbf{s}$ in (4.18) allows the global optimum of the approximate problem (4.16) to be found.*

Proof. The proof can be found in [12]. $\qquad\qquad\qquad\qquad\qquad\qquad\square$

Algorithm 4.1 Proposed JUM-QoS algorithm

Require: $s[1] \succ \mathbf{0}$ satisfying (4.24)–(4.25); $\gamma^{(i)\min} > 0, \forall i \in \mathscr{L}_m; \bar{\rho} \in [0, 1); a > 0; k > 1; \delta > 0; 0 < b < 1; \epsilon > 0; p[0] \succ \mathbf{0}$.

1: **while** $M/a \geq \epsilon$ **do**
2:     Set $t_s := 1$.
3:     **repeat**
4:         UE $i \in \mathscr{L}$ computes $v^{(i)}[t_s]$ by (4.23) and SINR target $\Gamma^{(i)}[t_s] = \log\left(\bar{\rho}s^{(i)}[t_s]/v^{(i)}[t_s]\right)$.
5:         Set $t_p := 0$.
6:         UE $i \in \mathscr{L}$ measures the actual SINR $\breve{\gamma}^{(i)}[t_s]$, and updates its power according to the Foschini-Miljanic's algorithm [18] until $p^{(i)}$ converges:

$$p^{(i)}[t_p + 1] := p^{(i)}[t_p] \frac{e^{\Gamma^{(i)}}}{\breve{\gamma}^{(i)}[t_s]}. \tag{4.22}$$

7:         UE $i \in \mathscr{L}$ measures interference $q^{(i)}[t_s]$, and finds $\Delta s^{(i)}[t_s]$ according to (4.19)–(4.20).
8:         Scale $\delta := b\,\delta$ until the resulting SINR target is strictly greater than $\gamma^{(i)\min}, \forall i \in \mathscr{L}_m$.
9:         UE $i \in \mathscr{L}$ updates $s^{(i)}[t_s + 1] := s^{(i)}[t_s] + \delta \Delta s^{(i)}[t_s]$.
10:       Set $t_s := t_s + 1$.
11:    **until** $s[t_s]$ converges to $\mathbf{s}^*$
12:     Set $s[1] := \mathbf{s}^*$, and update $a := k\,a$.
13: **end while**

4.2.2 Distributed Algorithm for Globally Maximized Joint Utility

We present in Algorithm 4.1 the Joint Utility Maximization with macrocell Quality-of-Service guarantee (JUM-QoS) algorithm to solve problem (4.10). Recall that resolving (4.16) only gives an approximate solution to problem (4.14), and in turn (4.10). Once problem (4.16) has been resolved, control parameter a needs to be regulated accordingly to improve the accuracy of approximation. Specifically, there are two levels of execution in Algorithm 4.1—the outer loop to update a, and the inner loop to find an optimal solution to the approximate problem (4.16). The resulting **s** of the current inner loop will be used in the next iteration of the outer loop.

Importantly enough, the proposed solution can be distributively implemented at each individual link with limited information being exchanged, either by means of broadcasting or over the available backhaul networks (e.g., DSL links). By assuming that channel gains between the downlink and the uplink are identical, and upon noticing that:

$$v^{(i)} = \sum_{j \in \mathscr{L}} H_{j,i} s^{(j)} = \sum_{j \neq i, \theta_j = \theta_i} s^{(j)} + \sum_{l \neq \theta_i} h_{l,i} \sum_{j, \theta_j = l} s^{(j)}, \tag{4.23}$$

the value of $v^{(i)}$ in Step 4 of Algorithm 4.1 can be computed and managed solely by UE $i \in \mathscr{L}$. Specifically, we require each BS l to broadcast the quantity $\sum_{j, \theta_j = l} s^{(j)}$ at a constant power. This permits UE i to also measure all the link gains $h_{l,i} = h_{i,l}$ required for the calculation of $\sum_{l \neq \theta_i} h_{l,i} \sum_{j, \theta_j = l} s^{(j)}$.

In Step 8 of Algorithm 4.1, to check the feasibility of the resulting target SINR $\tilde{\Gamma}^{(i)} = \log\left(\bar{\rho}\,\tilde{s}^{(i)}/\tilde{v}^{(i)}\right)$ associated with the search direction $\Delta s^{(i)}$, each UE $i \in \mathscr{L}$ computes $\tilde{s}^{(i)} := s^{(i)}[t_s] + \delta\Delta s^{(i)}[t_s]$, and subsequently $\tilde{v}^{(i)}$ as a function of $\tilde{s}$ [similar to (4.23)]. With channel gains $h_{l,i} = h_{i,l}$ already known, the computation of $\tilde{v}^{(i)}$ only requires $\tilde{s}^{(i)}$ to be exchanged, e.g., over backhaul links. It should also be noted that we need to initialize Algorithm 4.1 with a strictly feasible solution to ensure that $\Gamma^{(i)} > \Gamma^{(i)\min}$, $\forall i \in \mathscr{L}_m$. Since $\Gamma^{(i)} = \log\left(\bar{\rho}\,s^{(i)}/v^{(i)}\right)$ and $v^{(i)} = \sum_{i \in \mathscr{L}} H_{j,i}s^{(j)}$, this requirement corresponds to solving the following set of linear inequalities:

$$\bar{\rho}s^{(i)} - \sum_{j \in \mathscr{L}\setminus\{i\}} e^{\Gamma^{(i)\min}} H_{j,i}s^{(j)} > 0, \ \forall i \in \mathscr{L}_m \qquad (4.24)$$

$$s^{(i)} > 0, \ \forall i \in \mathscr{L}_f. \qquad (4.25)$$

Theorem 4.1. *As $\bar{\rho} \to 1$, the proposed JUM-QoS algorithm in Algorithm 4.1 converges to the global optimum of (4.10).*

Proof. The proof can be found in [12]. □

4.3 Distributed Power Control for Femtocell Utility Maximization and Macrocell SINR Balancing

JUM-QoS algorithm proposed in Algorithm 4.1 is applicable to a general scenario with the utilities of both macrocell and femtocell networks jointly optimized. In this case, it is noticed that the search space for Pareto-optimal SINRs always spans the whole $M + K$ dimensions. Consider a scenario in which MUEs do not require to maximize any utility, but rather only their predefined minimum SINRs be protected. A typical example is a macrocell network that mainly serves voice users coexisting with a data-serviced femtocell network. Specifically, this instance of problem corresponds to having $w_m = 0$ in the formulation (4.10). For notational convenience and without the loss of generality, let $w_f = 1$. Then, problem (4.10) is reduced to:

$$\max_{\boldsymbol{\Gamma} \in \mathbb{R}^{M+K}, \ \mathbf{p} \in \mathbb{R}_+^{M+K}} \sum_{i \in \mathscr{L}_f} U_i(\Gamma^{(i)})$$

$$\text{s.t. } \rho\left(\mathbf{HD}(e^{\boldsymbol{\Gamma}})\right) \leq \bar{\rho}, \qquad (4.26)$$

$$\Gamma^{(i)} \geq \Gamma^{(i)\min}, \ \forall i \in \mathscr{L}_m.$$

Upon observing the structure of the objective function in (4.26) and the monotonicity of SINR, it is shown in the sequel that the Pareto-optimal boundary of the feasible SINR region is confined to a smaller dimension. This property indeed gives rise to an algorithm that is more efficient than the general JUM-QoS solution.

4.3.1   Distributed Pareto-Optimal SINR Assignment

Let $\boldsymbol{\Gamma}_m^{\min} := \boldsymbol{\Gamma}^{\min}$, and perform the following matrix and vector partitions:
$$\mathbf{p} = \left[\mathbf{p}_m^T, \mathbf{p}_f^T\right]^T ; \mathbf{q} = \left[\mathbf{q}_m^T, \mathbf{q}_f^T\right]^T ; \boldsymbol{\Gamma} = \left[\boldsymbol{\Gamma}_m^T, \boldsymbol{\Gamma}_f^T\right]^T ; \boldsymbol{\sigma} = \left[\boldsymbol{\sigma}_m^T, \boldsymbol{\sigma}_f^T\right]^T ; \text{ and}$$
$$\mathbf{H} = \begin{bmatrix} \mathbf{H}_{11} & \mathbf{H}_{12} \\ \mathbf{H}_{21} & \mathbf{H}_{22} \end{bmatrix}, \text{ where } \mathbf{q}_m, \mathbf{p}_m, \boldsymbol{\sigma}_m \in \mathbb{R}_+^M, \boldsymbol{\Gamma}_m \in \mathbb{R}^M; \mathbf{q}_f, \mathbf{p}_f, \boldsymbol{\sigma}_f \in \mathbb{R}_+^K, \boldsymbol{\Gamma}_f \in$$
$\mathbb{R}^K; \mathbf{H}_{11} \in \mathbb{R}_+^{M \times M}, \mathbf{H}_{12} \in \mathbb{R}_+^{M \times K}, \mathbf{H}_{21} \in \mathbb{R}_+^{K \times M}, \text{ and } \mathbf{H}_{22} \in \mathbb{R}_+^{K \times K}.$

Proposition 4.3. *The optimal solution of (4.26) lies on the following boundary:*
$$\partial \mathscr{F}_{\bar{\rho}} := \{ \boldsymbol{\Gamma} = \left[\boldsymbol{\Gamma}_m; \boldsymbol{\Gamma}_f\right]; \boldsymbol{\Gamma}_m \in \mathbb{R}^M, \boldsymbol{\Gamma}_f \in \mathbb{R}^K$$
$$\text{s.t. } \rho\left(\mathbf{H}\mathbf{D}(e^{\boldsymbol{\Gamma}})\right) = \bar{\rho} \text{ and } \boldsymbol{\Gamma}_m = \boldsymbol{\Gamma}_m^{\min}\}. \tag{4.27}$$

Proof. The proof can be found in [12].   □

Proposition 4.3 implies that the search space for Pareto-optimal SINRs in this case is reduced to simply $\mathbb{R}^K$. An example of the above-derived boundary $\partial \mathscr{F}_{\bar{\rho}}$ is illustrated in Fig. 4.2, where the Pareto-optimal SINRs of a three-UE network lie within the curve connecting two points B and C. To unveil the complicated coupling between $\boldsymbol{\Gamma}_m$ and $\boldsymbol{\Gamma}_f$ in the relation $\rho\left(\mathbf{H}\mathbf{D}(e^{\boldsymbol{\Gamma}})\right) = \bar{\rho}$, the following result is now in order.

Proposition 4.4. *Suppose that we are operating on $\partial \mathscr{F}_{\bar{\rho}}$ and that $\rho\left(\mathbf{H}_{11}\mathbf{D}(e^{\boldsymbol{\Gamma}_m^{\min}})\right)$*
$< \bar{\rho}$. Then, $\rho\left(\mathbf{H}\mathbf{D}(e^{\boldsymbol{\Gamma}})\right) = \rho\left(\mathbf{F}\mathbf{D}(e^{\boldsymbol{\Gamma}_f})\right)$ holds, where $\mathbf{F}$ is a positive matrix defined as:
$$\mathbf{F} := \mathbf{H}_{21}\mathbf{D}(e^{\boldsymbol{\Gamma}_m^{\min}})\left[\bar{\rho}\mathbf{I}_M - \mathbf{H}_{11}\mathbf{D}(e^{\boldsymbol{\Gamma}_m^{\min}})\right]^{-1}\mathbf{H}_{12} + \mathbf{H}_{22}. \tag{4.28}$$

Proof. The proof can be found in [12].   □

Assumption $\rho\left(\mathbf{H}_{11}\mathbf{D}(e^{\boldsymbol{\Gamma}_m^{\min}})\right) < \bar{\rho}$ in Proposition 4.4 can be justified by first noting that channel matrix $\mathbf{H}$ is reduced to simply $\mathbf{H}_{11}$ if there is no femtocell deployed, and then applying the condition for the existence of a feasible power vector $\mathbf{p}_m = [p^{(1)}, \ldots, p^{(M)}]^T \succ \mathbf{0}$ in that case [16]. Essentially, Propositions 4.3 and 4.4 characterize the following Pareto-optimal SINR boundary of problem (4.26):
$$\partial \mathscr{F}_{\bar{\rho}} := \{ \boldsymbol{\Gamma} = \left[\boldsymbol{\Gamma}_m; \boldsymbol{\Gamma}_f\right]; \boldsymbol{\Gamma}_m \in \mathbb{R}^M, \boldsymbol{\Gamma}_f \in \mathbb{R}^K$$
$$\text{s.t. } \rho\left(\mathbf{F}\mathbf{D}(e^{\boldsymbol{\Gamma}_f})\right) = \bar{\rho} \text{ and } \boldsymbol{\Gamma}_m = \boldsymbol{\Gamma}_m^{\min}\}. \tag{4.29}$$

For every point on $\partial \mathscr{F}_{\bar{\rho}}$, it is impossible to increase the SINR of any one femto link without simultaneously reducing the SINR of some other femto links.

The finding of $\mathbf{F}$ in Proposition 4.4 also reveals that the performance of FUEs depends not only on the structure of the femtocell network (as reflected in $\mathbf{H}_{22}$), but also on the interaction between themselves with MUEs (as represented by $\mathbf{H}_{21}$ and $\mathbf{H}_{12}$). Moreover, the existence of $\mathbf{F}$ is conditional upon the particular values of $\mathbf{H}_{11}$ and $\boldsymbol{\Gamma}_m^{\min}$, i.e., $\rho\left(\mathbf{H}_{11}\mathbf{D}(e^{\boldsymbol{\Gamma}_m^{\min}})\right) < \bar{\rho}$. It is somewhat an expected result because MUEs have an absolutely higher priority in accessing radio resources. Such a condition also confirms that FUEs can attain their Pareto-optimal SINRs only if the performance of MUEs is, at least, unaffected.

The fact that matrix $\mathbf{F}$ is positive is critical, since it paves the way to adapt the load-spillage parametrization [9] to find all points on $\partial\mathscr{F}_{\bar{\rho}}$. Nevertheless, it is important to point out here that thanks to Propositions 4.3 and 4.4, one has to only deal with matrix $\mathbf{F}$ in a K-dimensional space instead of the original $(M + K) \times (M + K)$ channel matrix $\mathbf{H}$. Also note that $\mathbf{F}$ does not need to be primitive in the following result, unlike the strict condition on completely connected (i.e., primitive) matrices specifically required by [9].

Proposition 4.5. *For a fixed $\boldsymbol{\Gamma}_m^{\min}$, an SINR vector $\boldsymbol{\Gamma} = \left[\boldsymbol{\Gamma}_m; \boldsymbol{\Gamma}_f\right]$ lies on the boundary $\partial\mathscr{F}_{\bar{\rho}}$ defined in (4.29) if and only if there exist $\mathbf{s}_f \succ \mathbf{0}$ in $\mathbb{R}^K$ and $\bar{\rho} \in [0, 1)$ such that:*

$$\boldsymbol{\Gamma}_m = \boldsymbol{\Gamma}_m^{\min}, \tag{4.30}$$

$$\mathbf{s}_f^T\mathbf{F}\mathbf{D}(e^{\boldsymbol{\Gamma}_f}) = \bar{\rho}\mathbf{s}_f^T. \tag{4.31}$$

Proof. The proof can be found in [12].   □

Using Proposition 4.5, we can now parameterize all $\boldsymbol{\Gamma}_f$'s on the boundary $\partial\mathscr{F}_{\bar{\rho}}$ as follows. If we let:

$$\mathbf{v}_f := \mathbf{F}^T\mathbf{s}_f, \tag{4.32}$$

then (4.31) becomes $\mathbf{v}_f^T\mathbf{D}(e^{\boldsymbol{\Gamma}_f}) = \bar{\rho}\mathbf{s}_f^T$. From which,

$$\Gamma_f^{(i)} = \log\left(\bar{\rho}s_f^{(i)}/v_f^{(i)}\right); \; i = 1,\ldots,K. \tag{4.33}$$

After right-multiplying (4.31) by $\mathbf{F}$ and using (4.32), it is clear that $\mathbf{v}_f^T\mathbf{D}(e^{\boldsymbol{\Gamma}_f})\mathbf{F} = \bar{\rho}\mathbf{v}_f^T$, i.e., $\mathbf{v}_f$ is a left eigenvector associated with eigenvalue $\bar{\rho}$ of $\mathbf{D}(e^{\boldsymbol{\Gamma}_f})\mathbf{F}$. Furthermore, it is shown that $\bar{\rho} = \rho\left(\mathbf{D}(e^{\boldsymbol{\Gamma}_f})\mathbf{F}\right) = \rho\left(\mathbf{F}\mathbf{D}(e^{\boldsymbol{\Gamma}_f})\right)$ [14, Theorem 1.3.20].

Once $s_f^{(i)}$ is known, the computation of $\Gamma_f^{(i)}$ in (4.33) requires $v_f^{(i)}$ to be found by (4.32). However, as $\mathbf{F}$ involves a matrix inverse operation [see (4.28)], it is not yet straightforward to find $v_f^{(i)}$ distributively. Using (4.28), we rewrite (4.32) as:

$$\mathbf{v}_f^T = \mathbf{s}_f^T \mathbf{H}_{21} \mathbf{D}(e^{\boldsymbol{\Gamma}_m^{\min}}) \left[\bar{\rho} \mathbf{I}_M - \mathbf{H}_{11} \mathbf{D}(e^{\boldsymbol{\Gamma}_m^{\min}}) \right]^{-1} \mathbf{H}_{12} + \mathbf{s}_f^T \mathbf{H}_{22}, \qquad (4.34)$$

and define $\mathbf{s}_m \in \mathbb{R}_+^M$ as

$$\mathbf{s}_m^T := \mathbf{s}_f^T \mathbf{H}_{21} \mathbf{D}(e^{\boldsymbol{\Gamma}_m^{\min}}) \left[\bar{\rho} \mathbf{I}_M - \mathbf{H}_{11} \mathbf{D}(e^{\boldsymbol{\Gamma}_m^{\min}}) \right]^{-1}. \qquad (4.35)$$

Proposition 4.6. *Given an initialization* $\mathbf{s}_m^T[0] \succ \mathbf{0}$, $\mathbf{s}_m^T$ *can be realized by the following update:*

$$\mathbf{s}_m^T[t+1] = \frac{1}{\bar{\rho}} \mathbf{s}_m^T[t] \mathbf{H}_{11} \mathbf{D}(e^{\boldsymbol{\Gamma}_m^{\min}}) + \frac{1}{\bar{\rho}} \mathbf{s}_f^T \mathbf{H}_{21} \mathbf{D}(e^{\boldsymbol{\Gamma}_m^{\min}}). \qquad (4.36)$$

Proof. The proof can be found in [12]. □

Notice that the i-th component of $\mathbf{s}_m[t+1]$ in (4.36) is actually:

$$s_m^{(i)}[t+1] = \frac{e^{\Gamma_m^{\min(i)}}}{\bar{\rho}} \left[\sum_{j=1}^M H_{11}^{(j,i)} s_m^{(j)}[t] + \sum_{j=1}^K H_{21}^{(j,i)} s_f^{(j)} \right], \quad i = 1, \dots, M. \qquad (4.37)$$

From (4.1) and upon recalling the partition of $\mathbf{H}$, (4.37) is indeed:

$$
\begin{aligned}
s_m^{(i)}[t+1] &= \frac{e^{\Gamma_m^{\min(i)}}}{\bar{\rho}} \left[\sum_{j \in \mathcal{L}_m \setminus \{i\}} s_m^{(j)}[t] + \sum_{j \in \mathcal{L}_f} h_{\theta_j,i} s_f^{(j-K)} \right] \\
&= \frac{e^{\Gamma_m^{\min(i)}}}{\bar{\rho}} \left[\sum_{j \in \mathcal{L}_m \setminus \{i\}} s_m^{(j)}[t] + \sum_{l \neq \theta} h_{l,i} \sum_{j,\theta_j = l} s_f^{(j-K)} \right], \quad i = 1, \dots, M.
\end{aligned}
\qquad (4.38)
$$

Clearly, $s_m^{(i)}[t+1]$ consists of an *internal* component $\sum_{j \in \mathcal{L}_m \setminus \{i\}} s_m^{(j)}[t]$ due to other MUEs, and an *external* component $\sum_{l \neq \theta} h_{l,i} \sum_{j,\theta_j = l} s_f^{(j-K)}$ due to all FUEs (with θ denoting the macrocell BS).

With $\mathbf{s}_f \succ \mathbf{0}$ known and once $\mathbf{s}_m \succ \mathbf{0}$ has been determined, $\mathbf{v}_f$ can be readily computed. From (4.34) and (4.35), $\mathbf{v}_f^T = \mathbf{s}_m^T \mathbf{H}_{12} + \mathbf{s}_f^T \mathbf{H}_{22}$. Then, its component can be found according to:

$$v_f^{(i)} = h_{\theta,i} \sum_{j \in \mathcal{L}_m} s_m^{(j)} + \sum_{j \neq i, \theta_j = \theta_i} s_f^{(j)} + \sum_{l \neq \theta_i} h_{l,i} \sum_{j,\theta_j = l} s_f^{(j)}, \quad i = 1, \dots, K. \quad (4.39)$$

It is worth commenting on that the first term of (4.39) amounts to the effects from all MUEs, whereas the second term from the FUEs within the same femtocell, and the third term from the FUEs in all other femtocells.

4.3.2   Distributed Algorithm for Femtocell Utility Maximization and Macrocell SINR Balancing

The above parametrization $\boldsymbol{\Gamma}_f = \boldsymbol{\Gamma}_f(\mathbf{s}_f, \boldsymbol{\Gamma}_m^{\min}, \bar{\rho})$ allows us to find all points on $\mathscr{F}_{\bar{\rho}}$. By fixing $\bar{\rho} \in [0, 1)$ and upon applying that parametrization, (4.26) can be solved via an equivalent optimization problem, albeit in the new variable $\mathbf{s}_f$. The latter involves finding a direction of $\mathbf{s}_f$ that leads $\boldsymbol{\Gamma}_f$ and $\mathbf{p}$ to the optimum of the original problem. With a scalar step size $\delta_f > 0$, we propose to update $s_f^{(i)}$ as:

$$s_f^{(i)}[t+1] := s_f^{(i)}[t] + \delta_f \, \Delta s_f^{(i)}[t+1], \tag{4.40}$$

$$\Delta s_f^{(i)}[t+1] := U_i'\left(\Gamma_f^{(i)}\right) / (\bar{\rho} q_f^{(i)}) - s_f^{(i)}[t], \tag{4.41}$$

for $i = 1, \ldots, K$. Upon recalling that $\mathbf{s}_f$ is a left eigenvector associated with eigenvalue $\bar{\rho}$ of $\mathbf{FD}(e^{\boldsymbol{\Gamma}_f})$, it can be proven that the update of $\mathbf{s}_f$ in (4.40) and (4.41) actually represents an ascent direction for $U(\mathbf{s}_f) := \sum_{i \in \mathscr{L}_f} U_i(\Gamma^{(i)})$ [9, Appendix D]. We also note that the update of MUE load $s_m^{(i)}$ in (4.38) is totally different from that of FUE load $s_f^{(i)}$ in (4.40).

We present in Algorithm 4.2 the Femtocell Utility Maximization with Macrocell SINR Balancing (FUM-MSB) algorithm. Again, it is assumed that channel gains between the downlink and the uplink are identical. Here, $s_m^{(i)}[t+1]$ in Step 4 of Algorithm 4.2 is computed and managed by MUE $i \in \mathscr{L}_m$. Specifically, each femtocell BS l is required to broadcast $\sum_{j, \theta_j = l} s_f^{(j)}$ at a constant power. This allows MUE i to also measure all channel gains $h_{l,i} = h_{i,l}$ required for the calculation of $\sum_{l \neq \theta_i} h_{l,i} \sum_{j, \theta_j = l} s_f^{(j)}$. On the other hand, macrocell BS communicates the quantity $\sum_{j \in \mathscr{L}_m} s_m^{(j)}[t]$ to all MUEs, which then permits MUE i to easily compute $\sum_{j \in \mathscr{L}_m \setminus \{i\}} s_m^{(j)}[t] = \sum_{j \in \mathscr{L}_m} s_m^{(j)}[t] - s_m^{(i)}[t]$. Finally, MUE i reports the resulting $s_m^{(i)}[t+1]$ back to macrocell BS for the computation of $\mathbf{s}_m$ in the next iteration. Note that each femtocell BS l only needs to broadcast $\sum_{j, \theta_j = l} s_f^{(j)}$ once. As well, $\sum_{j \in \mathscr{L}_m} s_m^{(j)}[t]$ and $s_m^{(i)}[t+1]$ can be exchanged locally between macrocell BS and MUE i over the control channel of link i.

The computation of $v_f^{(j)}$ in Step 7 of Algorithm 4.2 can also be done by FUE j. Once $\mathbf{s}_m$ has been determined (i.e., its update (4.38) has converged), macrocell BS broadcasts the quantity $\sum_{i \in \mathscr{L}_m} s_m^{(i)}$, again at a constant power. Recall that all summations $\sum_{i, \theta_i = l} s_f^{(i)}$ have already been received at FUE j from

---

Algorithm 4.2 Proposed FUM-MSB algorithm

Require: $\gamma_m^{\min} \succ 0, \bar{\rho} \in [0, 1)$, and $\delta_f > 0$.

1: Initialize: $\mathbf{p}_m[0] \succ 0, \mathbf{p}_f[0] \succ 0, \mathbf{s}_f[0] \succ 0; t_m := 1, t_f := 1$.

2: Set arbitrary $\mathbf{s}_m[0] \succ 0$.

3: **repeat**

4:      MUE i computes $s_m^{(i)}[t_m]$ by (4.38).

5:      Set $t_m := t_m + 1$.

6: **until** $\mathbf{s}_m$ converges

7: FUE j computes $v_f^{(j)}$ by (4.39) and SINR target $\Gamma_f^{(j)}$ by (4.33).

8: Set $t_p := 0$.

9: MUE i measures the actual SINR $\breve{\gamma}^{(i)}$, and updates its power by Foschini-Miljanic's algorithm [18], i.e., $p_m^{(i)}[t_p + 1] := p_m^{(i)}[t_p]\gamma_m^{(i)\min}/\breve{\gamma}^{(i)}$ until $p_m^{(i)}$ converges.

10: FUE j measures the actual SINR $\breve{\gamma}^{(j)}$, and updates its power as $p_f^{(j)}[t_p + 1] := p_f^{(j)}[t_p]e^{\Gamma_f^{(j)}}/\breve{\gamma}^{(j)}$ until $p_f^{(j)}$ converges.

11: FUE j measures interference $q_f^{(j)}$.

12: FUE j updates $s_f^{(j)}[t_f + 1]$ according to (4.40)–(4.41).

13: Set $t_m := 1, t_f := t_f + 1$, go back to Step 2 and repeat until $\mathbf{s}_f$ converges.

---

all femtocell BSs l (including the one that serves FUE j). Together with the assumption of symmetric downlink-uplink channel gains, $v_f^{(j)}$ can thus be computed according to (4.39). As well, the update of $s_f^{(j)}$ in Step 12 of Algorithm 4.2 can be accomplished in a completely distributed manner by FUE j with only local information required. Over its own control channel, FUE j then reports the new value of $s_f^{(j)}$ to its servicing femtocell BS θ_j, to be used in the next iteration.

Theorem 4.2. *For a sufficiently small $\delta_f > 0$ and as $\bar{\rho} \to 1$, the proposed FUM-MSB algorithm in Algorithm 4.2 converges to the globally optimal solution of (4.26).*

Proof. The proof can be found in [12]. □

4.3.3 Advantages of FUM-MSB Algorithm

Although both proposed schemes can be used to solve the same problem (4.26), FUM-MSB algorithm [see Algorithm 4.2] outperforms JUM-QoS algorithm [see Algorithm 4.1] in this specific case. Compared with JUM-QoS, FUM-MSB solution converges more quickly to the optimal points. This is because the latter operates independently of the total number of MUEs M, and its search space for Pareto-optimal SINR is simply confined to $\mathbb{R}^K$. In practical networks with a large number of MUEs, FUM-MSB algorithm is thus more scalable. Moreover, this solution offers a substantial reduction in computational complexity. Recall that JUM-QoS algorithm is based on the update of penalty factor a in another time scale, and hence its performance can be sensitive to the actual values of a. On the contrary, penalty method is not needed at all in FUM-MSB scheme. Further, this specific algorithm does not search for a feasible starting point [similar to that in (4.24) and (4.25)], which involves some message exchange among all macro and femto BSs. Given that only limited backhaul network capacity is available for femtocells, this kind of message passing can be a performance bottleneck in certain scenarios. Note also that the scaling of step size δ in Step 8 of JUM-QoS algorithm might lead to the ripple effect around the optimum, as one tries to push Γ into the strict interior of the feasible set. This instability issue does not happen to FUM-MSB solution.

4.4 Illustrative Results

We present numerical results to illustrate the performance of the two proposed algorithms—JUM-QoS in Algorithm 4.1 and FUM-MSB in Algorithm 4.2. The network setting and user placement in our examples are shown in Fig. 4.3, where MUEs and FUEs are randomly deployed inside circles of radii of 500 m and 50 m, respectively. In particular, we assume there are $M = 10$ MUEs, whereas $K = 20$ FUEs are divided equally among 4 femtocells (i.e., 5 FUEs per femtocell). The uplink case is considered in all simulations. The absolute channel gain from UE j to BS θ_i is calculated according to:

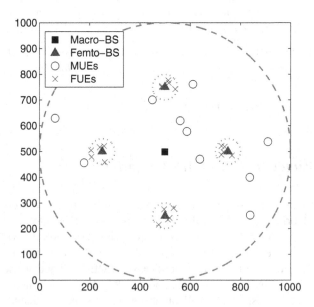

Fig. 4.3 Network setup for the numerical examples

Fig. 4.4 Convergence process of JUM-QoS algorithm (Algorithm 4.1)

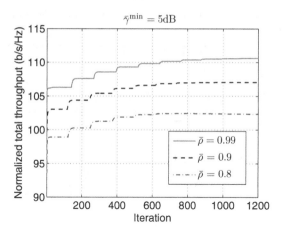

$$
\bar{h}_{\theta_i,j} = \begin{cases} d_{\theta_i,j}^{-\beta}, & \text{if } \theta_j = \theta_i, \\ d_{\theta_i,j}^{-\beta}/(10^{\bar{\kappa}/10}), & \text{if } \theta_j \neq \theta_i, \end{cases} \tag{4.42}
$$

where $d_{\theta_i,j}$ is their corresponding geographical distance, pathloss exponent $\beta = 3$ is used, and $\bar{\kappa} = 10\,\text{dB}$ is chosen to represent the extra cross-cell signal loss due to penetration through walls (as FUEs are typically deployed indoors).

For simplicity, we consider unit bandwidth. The throughput, normalized over the total bandwidth, is thus expressed in terms of bps/Hz. Gaussian noise power is taken as $\sigma^{(i)} = 10^{-6}$, $\forall i \in \mathcal{L}$. Normalized minimum SINRs $\gamma_m^{\min} = [\gamma^{(1)\,\min}, \ldots, \gamma^{(M)\,\min}]^T$, are assumed equal for all MUEs, i.e., $\gamma^{(i)\,\min} = \gamma^{\min}$, $\forall i \in \mathcal{L}_m$, chosen such that $\rho\left(\mathbf{HD}([\gamma_m^{\min}; \mathbf{0}_K])\right) \leq \bar{\rho} < 1$. Because all the results obtained in the previous sections are applied to the normalized SINR $\gamma^{(i)}$, $\forall i \in \mathcal{L}$, the actual attained SINR in the numerical examples must be recovered according to $\bar{\gamma}^{(i)} = G\gamma^{(i)}$, $\forall i \in \mathcal{L}$, where G is the processing gain. While it is possible to select other values of G, we choose $G = 32$ for this particular network realization so that the actual minimum SINR $\bar{\gamma}^{\min} = G\gamma^{\min}$ is within a practical range (i.e., from 5 dB to almost 8.5 dB). Unless stated otherwise, 3-fair utility function is used, i.e., $\alpha = 3$ in (4.11). We set the error tolerance for the convergence of the proposed schemes and Foschini-Miljanic's algorithm as $\epsilon = 10^{-4}$ and $\epsilon_p = 10^{-10}$, respectively. For JUM-QoS algorithm, $a = 2, k = 2, \delta = 0.1, b = 0.8$ are assumed, whereas for FUM-MSB algorithm $\delta_f = 0.1$.

Figure 4.4 demonstrates the convergence of JUM-QoS algorithm for $w_m = w_f = 0.5$. At each stage of the proposed log-barrier penalization (which corresponds to a given value of a), the algorithm quickly converges in some tens of iterations, and an improvement in the total throughput is observed. After several updates of a, the final convergence is realized. As $\bar{\rho}$ tends to 1, the total sum rates of all MUEs and FUEs increase. This is because the feasible SINR region becomes larger as $\bar{\rho}$ grows, meaning that more capacity is available. Note also that JUM-QoS

Table 4.1 Performance of JUM-QoS algorithm (Algorithm 4.1) with $\bar{\rho} = 0.99$

$\bar{\gamma}^{\min}$ (dB)	6	7	8	8.4	8.45	8.475
$\rho\left(\mathbf{HD}\left([\boldsymbol{\gamma}_m^{\min}, \mathbf{0}_K]\right)\right)$	0.5598	0.7048	0.8873	0.9729	0.9842	0.9898
$\max\{\bar{\boldsymbol{\gamma}}_m\}$ (dB)	8.4331	8.4331	8.4331	8.4895	8.4688	8.4754
$\min\{\bar{\boldsymbol{\gamma}}_m\}$ (dB)	8.3637	8.3637	8.3637	**8.4000**	**8.4500**	**8.4750**
Total r_{MUE} (bps/Hz)	29.8930	29.8930	29.8930	29.8955	30.0110	30.0702
Total r_{FUE} (bps/Hz)	80.7477	80.7477	80.7477	80.5960	74.7508	37.5895

Fig. 4.5 Throughput tradeoff between macrocell and femtocell networks by JUM-QoS algorithm (Algorithm 4.1)

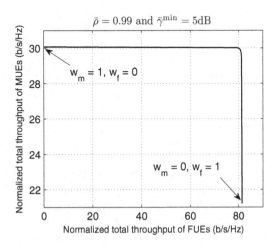

is expected to approach the global optimum in the limit $\bar{\rho} \to 1$ [see Proposition 4.2 and Theorem 4.1].

On the other hand, the results presented in Table 4.1 can be interpreted as follows. Without any MUEs' prescribed minimum SINR $\bar{\gamma}_m^{\min}$, the optimal SINR assignment for all MUEs that maximizes the joint utilities of both macrocell and femtocells can be denoted as $\check{\boldsymbol{\gamma}}_m^*$, whose entries range from 8.3637 to 8.4331 dB. As such, including any SINR $\bar{\gamma}_m^{\min} \preceq \check{\boldsymbol{\gamma}}_m^*$ in the constraint set does not change this final solution. While different network configurations correspond to different specific values of $\check{\boldsymbol{\gamma}}_m^*$, this solution $\check{\boldsymbol{\gamma}}_m^*$ will no longer be feasible for any constraint $\bar{\gamma}_m^{\min} \succ \check{\boldsymbol{\gamma}}_m^*$. Remarkably, the proposed JUM-QoS algorithm always guarantees that the resulting SINRs of all MUEs are actually greater than $\bar{\gamma}_m^{\min}$ in that case, as evidenced in the last three columns of Table 4.1. Furthermore, it is noteworthy that a small variation in $\bar{\gamma}_m^{\min}$ in this range of SINR may significantly reduce the remaining network capacity available for FUEs. From Table 4.1, as the prioritized MUEs demand for a slight increase of 0.025 dB in $\bar{\gamma}_m^{\min}$, the total throughput of all femtocells is decreased by half, dropping from almost 75 bps/Hz to about 37.5 bps/Hz.

To flexibly share the radio resources among MUEs and FUEs, the general JUM-QoSalgorithm can designate the importance toward either macrocell or femtocell network by varying the values of w_m and w_f. In Fig. 4.5, the achieved throughput of both networks is displayed for $w_m = 0 : 0.1 : 1$ and $w_f = 1 - w_m$.

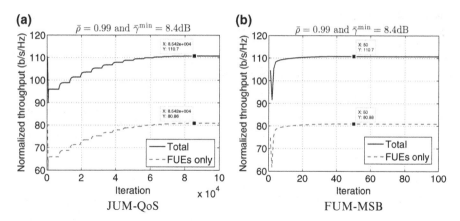

Fig. 4.6 Performance comparison of JUM-QoS (Algorithm 4.1) and FUM-MSB (Algorithm 4.2) algorithms (**a**) JUM-QoS (**b**) FUM-MSB

Clearly, by changing from $w_m = 1$ to $w_m = 0$, i.e., MUEs only require to have their minimum QoS maintained rather than their utility solely optimized, the throughput improvement in the femtocell network is ninefold the amount of rate loss in the macrocell. Such a pronounced gain can be explained by the fact that FUEs are located in close proximity to their corresponding BSs, and thus are able to achieve potentially much higher data rates compared to MUEs.

To compare its performance with that of the specific FUM-MSB algorithm, we set $w_m = 0, w_f = 1$ in the general JUM-QoS algorithm. With $\bar{\rho} = 0.99$ and $\bar{\gamma}_m^{\min} = 8.4$ dB, this general algorithm takes almost 10^5 iterations to reach the final optimal solution [see Fig. 4.6a]. The main reason for such a long converging time is that it takes the log-barrier penalty method quite a lot of efforts to push the MUEs' SINRs to be very close to the boundary of the feasible SINR region, i.e., to achieve $\bar{\gamma}_m^{\min}$. Even so, since JUM-QoS algorithm operates strictly inside the feasible region, macrocell SINR targets can never get exactly equal to 8.4 dB. On the contrary, FUM-MSB algorithm settles very quickly to the global optimum in as few as 10 iterations [see Fig. 4.6b], with the *exact* SINR 8.4 dB obtained for all MUEs. The latter fact also explains why the total femtocell throughput given by this algorithm is somewhat greater than that by the JUM-QoS counterpart. Moreover, computational results suggest that FUM-MSB algorithm does not experience any ripple effect that occurs to JUM-QoS scheme around the optimum point.

The issue of fairly utilizing the available radio resources can be effectively resolved by regulating α in the utility function. Figure 4.7 shows the minimum and maximum throughput among all the femtocells, given by FUM-MSB solution for different values of α. Apparently, as α increases, the FUE whose data rate is the highest (likely due to its advantageous link conditions) sees a decline in its

Fig. 4.7 Fairness achieved by the use of different utility functions in FUM-MSB algorithm (Algorithm 4.2)

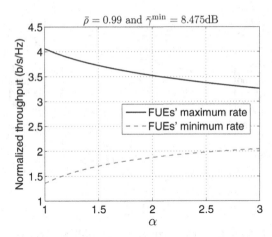

throughput, whereas the FUE with the lowest throughput has its data rate gradually enhanced. For a sufficiently large value of α, the FUEs' minimum throughput is expected to reach a plateau, meaning that max-min fairness is realized at that point.

References

1. D. T. Ngo, L. B. Le, T. Le-Ngoc, E. Hossain, and D. I. Kim, "Distributed interference management in femtocell networks," in *Proc. IEEE Veh. Tech. Conf. (VTC-Fall)*, San Franciso, CA, Sep. 2011, pp. 1–5.
2. D. T. Ngo, L. B. Le, T. Le-Ngoc, E. Hossain, and D. I. Kim, "Distributed interference management in two-tier CDMA femtocell networks," *IEEE Trans. Wireless Commun.*, vol. 11, no. 3, pp. 979–989, Mar. 2012.
3. C. U. Saraydar, N. B. Mandayam, and D. J. Goodman, "Pricing and power control in a multicell wireless data network," *IEEE J. Select. Areas Commun.*, vol. 19, no. 10, pp. 1883–1892, Oct. 2001.
4. C. U. Saraydar, N. B. Mandayam, and D. J. Goodman, "Efficient power control via pricing in wireless data networks," *IEEE Trans. Commun.*, vol. 50, no. 2, pp. 291–303, Feb. 2002.
5. M. Xiao, N. B. Shroff, and E. K. P. Chong, "A utility-based power control scheme in wireless cellular systems," *IEEE/ACM Trans. Netw.*, vol. 11, no. 2, pp. 210–221, Apr. 2003.
6. S. Koskie and Z. Gajic, "A Nash game algorithm for SIR-based power control in 3G wireless CDMA networks," *IEEE/ACM Trans. Netw.*, vol. 13, no. 5, pp. 1017–1026, Oct. 2005.
7. V. Chandrasekhar, J. G. Andrews, T. Muharemovic, and Z. Shen, "Power control in two-tier femtocell networks," *IEEE Trans. Wireless Commun.*, vol. 8, no. 8, pp. 4316–4328, Aug. 2009.
8. M. Rasti, A. R. Sharafat, and B. Seyfe, "Pareto-efficient and goal-driven power control in wireless networks: A game-theoretic approach with a novel pricing scheme," *IEEE/ACM Trans. Netw.*, vol. 17, no. 2, pp. 556–569, Apr. 2009.
9. P. Hande, S. Rangan, M. Chiang, and X. Wu, "Distributed uplink power control for optimal SIR assignment in cellular data networks," *IEEE/ACM Trans. Netw.*, vol. 16, no. 6, pp. 1420–1433, Dec. 2008.
10. D. T. Ngo, L. B. Le, and T. Le-Ngoc, "Distributed Pareto-optimal power control in femtocell networks," in *Proc. IEEE Intl. Symp. on Personal, Indoor and Mobile Radio Commun. (PIMRC)*, Toronto, ON, Canada, Sep. 2011, pp. 222–226.

11. D. T. Ngo, L. B. Le, and T. Le-Ngoc, "Joint utility maximization in two-tier networks by distributed Pareto-optimal power control," in *Proc. IEEE Vehicular Technology Conf. (VTC-Fall)*, Quebec City, QC, Canada, Sep. 2012, pp. 1–5.

12. D. T. Ngo, L. B. Le, and T. Le-Ngoc, "Distributed Pareto-optimal power control for utility maximization in femtocell networks," *IEEE Trans. Wireless Commun.*, vol. 11, no. 10, pp. 3434–3446, Oct. 2012.

13. N. Bambos, S. C. Chen, and G. J. Pottie, "Channel access algorithms with active link protection for wireless communication networks with power control," *IEEE/ACM Trans. Netw.*, vol. 8, no. 5, pp. 583–597, Oct. 2000.

14. A. Horn and A. Johnson, *Matrix Analysis*, 1st ed. Cambridge University Press, 1985.

15. J. Mo and J. Walrand, "Fair end-to-end window-based congestion control," *IEEE/ACM Trans. Netw.*, vol. 8, no. 5, pp. 556–567, Oct. 2000.

16. J. Zander, "Performance of optimum transmitter power control in cellular radio systems," *IEEE Trans. Veh. Technol.*, vol. 41, no. 1, pp. 57–62, Feb. 1992.

17. H. Boche and S. Stanczak, "Convexity of some feasible QoS regions and asymptotic behavior of the minimum total power in CDMA systems," *IEEE Trans. Commun.*, vol. 52, no. 12, pp. 2190–2197, Dec. 2004.

18. G. J. Foschini and Z. Miljanic, "A simple distributed autonomous power control algorithm and its convergence," *IEEE Trans. Veh. Technol.*, vol. 42, no. 4, pp. 641–646, Nov. 1993.

Chapter 5
Joint Power and Subchannel Allocation in Heterogeneous OFDMA Small-Cell Networks

Compared to the orthogonal deployment, cochannel deployment is more attractive in heterogeneous network settings as it can offer a much higher spectral efficiency [1–3]. Chapters 3 and 4 have presented power control algorithms for cochannel interference management in CDMA-based heterogeneous networks [4–8]. On the other hand, because of its flexibility in allocating the radio spectrum, OFDMA has been used as the air-interface technology in LTE femtocells, i.e., home evolved Node Bs (HeNBs) [9, 10]. With OFDMA, intracell interference is eliminated thanks to the exclusive channel assignment, in which a subchannel is allotted to at most one UE in each cell at any given time.

Nonetheless, cochannel deployment implies that an OFDM subchannel can be shared by UEs from different cells, giving rise to the intercell interference (ICI). Furthermore, there is another source of technical difficulty here—the subchannel assignment problem which involves the allocation of radio frequencies to different UEs in multiple cells. To directly solve this combinatorial problem, approaches with an exponential complexity are required. Therefore, the successful development of any resource allocation scheme for OFDMA-based heterogeneous small-cell networks relies upon how one can effectively overcome such a difficult problem.

This chapter provides a new formulation for the downlink subchannel and power allocation problem in an OFDMA-based small-cell network, where the network capacity of the prioritized macrocell is protected regardless of any femtocell deployment [11, 12]. Because the data rate function is highly nonconvex in this multiuser multicell setting, it is especially challenging to manage the ICI due to radio spectrum being shared among UEs from different cells. Such a critical issue is further complicated by the aforementioned nontrivial task of assigning multiple OFDM subchannels to individual UEs in each cell so as to maximize the total femtocell throughput. Moreover, the strict requirement of satisfying a minimum data rate constraint for the macrocell at all times presents another dimension of difficulty in finding an optimal solution for resource allocation.

Towards solving this multi-dimensional problem, an iterative algorithm is developed that alternatively assigns OFDM subchannels to UEs and allocates transmit

D.T. Ngo and T. Le-Ngoc, *Architectures of Small-Cell Networks and Interference Management*, SpringerBriefs in Computer Science, DOI 10.1007/978-3-319-04822-2__5,
© The Author(s) 2014

power to BSs. For each subchannel assignment subproblem, the optimal strategy is that every cell gives subchannels to the serviced UEs with the highest corresponding SINRs. For the power allocation subproblems, the successive convex approximation (SCA) approach [13] is employed, in which the original highly nonconvex problem is transformed into a series of relaxed convex programs. With the arithmetic-geometric mean (AGM) approximation, a sequence of geometric programs [14] is solved after condensing a posynomial into a monomial. In the logarithmic approximation, upon lower-bounding the highly nonconcave rate function by a concave function [15], one has to simply deal with standard concave maximization problems. In the difference-of-two-concave-functions (D.C.) approximation, the data rate is represented as a difference of two concave functions [16, 17] and a set of improved feasible solutions is then generated.

The SCA approach has been applied to various scenarios, e.g., single-carrier systems [18, 19], digital subscriber lines (DSL) [15], and multi-carrier homogeneous networks [20]. However, the network setting considered here is very different, making the existing solutions not directly applicable. Specifically, the SCA approach is adapted to systematically address the critical issue of interference in dense small-cell OFDMA-based heterogeneous networks with diverse user QoS constraints. It is established that all three power optimization policies based on the SCA approach eventually converge to optimal transmit power solutions for any given subchannel allotment. Also shown is that the overall joint subchannel and power allocation algorithm is convergent while ensuring that the macrocell sum rate is always above a prescribed value. For the AGM approximation, it is proposed that the network optimization is performed by a central processing unit, e.g., at an operation, administration and management (OAM) server. For the logarithmic and D.C. approximations, distributed implementations of the proposed algorithm are provided, wherein individual BSs compute the optimal OFDM subchannel and power allocation for their own servicing cells.

5.1 System Model and Problem Formulation

We consider the downlink of a two-tier wireless network, in which M_f newly deployed femtocells share the available radio spectrum with one existing macrocell. We assume that each cell is serviced by one BS and denote the set of all BSs as $\mathcal{M} = \{0, 1, 2, \ldots, M_f\}$. For simplicity, cell m means the cell served by BS m. Without loss of generality, the macrocell BS is indexed by 0 and femtocell BSs by $1, 2, \ldots, M_f$. If we denote the set of all UEs associated with BS $m \in \mathcal{M}$ as $\mathcal{K}_m$, then the number of UEs in cell m is $K_m = |\mathcal{K}_m|$. The total number of UEs in the entire network is thus $K = \sum_{m \in \mathcal{M}} K_m$. We also make the reasonable assumption that the association of a certain UE with its own BS is fixed during the runtime of the underlying network optimization process. A typical example of the two-tier wireless network under investigation is illustrated in Fig. 5.1.

Fig. 5.1 Example of a two-tier OFDMA wireless heterogeneous network

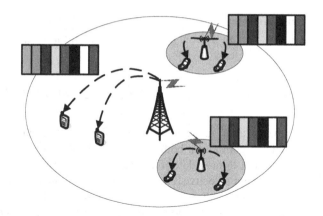

Denote the set of all accessible frequencies as $\mathcal{N}$, which consists of $|\mathcal{N}| = N$ OFDM subchannels. With OFDMA, we enforce an exclusive channel assignment in each cell, i.e., a particular OFDM subchannel can be used by at most one UE in that cell at a given time. To achieve a higher spectral efficiency, any UE—be it an MUE or an FUE—from different cells is allowed to share the same subchannel via cochannel deployment. In this paper, we assume that the time scale of network topology change is very small compared to that of power adaptation and subchannel allocation. The time scale of data transmission is assumed to be much shorter than that of the underlying optimization, allowing for any short-term statistical variations to be averaged out.

Let $h_{m,k}^{(n)}$ be the channel gain from BS $m \in \mathcal{M}$ to UE k on subchannel $n \in \mathcal{N}$, and $p_m^{(n)}$ the transmit power of $m \in \mathcal{M}$ on $n \in \mathcal{N}$. Denote $\mathbf{p}^{(n)} = [p_0^{(n)}, p_1^{(n)}, \ldots, p_{M_f}^{(n)}]^T$, $\mathbf{p}_m = [p_m^{(1)}, p_m^{(2)}, \ldots, p_m^{(N)}]^T$, and $\mathbf{p} = \text{vec}[\mathbf{p}_0, \mathbf{p}_1, \cdots, \mathbf{p}_{M_f}]$. The received SINR at UE $k \in \mathcal{K}_m$ on subchannel $n \in \mathcal{N}$ is expressed as:

$$\gamma_{m,k}^{(n)}(\mathbf{p}^{(n)}) = \frac{h_{m,k}^{(n)} p_m^{(n)}}{\sum_{j \in \mathcal{M} \setminus \{m\}} h_{j,k}^{(n)} p_j^{(n)} + \sigma_k^{(n)}}, \tag{5.1}$$

where $\sigma_k^{(n)}$ is the noise power at the receiver of UE k on n. Recall that there is no intracell interference within each cell due to the assumption of OFDMA. The instantaneous data rate attained by UE $k \in \mathcal{K}_m$ on subchannel n can then be written as:

$$r_{m,k}^{(n)}(\mathbf{p}^{(n)}) = \ln\left(1 + \gamma_{m,k}^{(n)}(\mathbf{p}^{(n)})\right). \tag{5.2}$$

Denote the subchannel assignment by a binary variable $\rho_{m,k}^{(n)}$, where $\rho_{m,k}^{(n)} = 1$ if $n \in \mathcal{N}$ is assigned to $k \in \mathcal{K}_m$ and $\rho_{m,k}^{(n)} = 0$ otherwise. For notational convenience, let us also denote $\boldsymbol{\rho}_{m,k} = [\rho_{m,k}^{(1)}, \rho_{m,k}^{(2)}, \ldots, \rho_{m,k}^{(N)}]^T$, $\boldsymbol{\rho}_m = \text{vec}(\boldsymbol{\rho}_{m,1}, \boldsymbol{\rho}_{m,2}, \ldots, \boldsymbol{\rho}_{m,K_m})$, and $\boldsymbol{\rho} = \text{vec}(\boldsymbol{\rho}_0, \boldsymbol{\rho}_1, \ldots, \boldsymbol{\rho}_M)$. Then, the total throughput attained by UE $k \in \mathscr{K}_m$ can be computed as:

$$r_{m,k}(\boldsymbol{\rho}_{m,k}, \mathbf{p}) = \sum_{n \in \mathscr{N}} \rho_{m,k}^{(n)} r_{m,k}^{(n)}(\mathbf{p}^{(n)}). \tag{5.3}$$

This paper aims to devise an optimal joint subchannel assignment and power allocation algorithm for both MUEs and FUEs. In a heterogeneous network setting, it is essential to provide differentiated QoS so that each class of UEs can achieve its own design objective. Here, the lower-tier femtocells target at maximizing their total throughput, whereas the prioritized macrocell specifically demands that their existing network capacity not be reduced regardless of any femtocell deployment. The radio resource management problem can thus be formulated as follows.

$$\max_{\mathbf{p}, \rho} \sum_{m \in \mathscr{M} \setminus \{0\}} \sum_{k \in \mathscr{K}_m} \sum_{n \in \mathscr{N}} \rho_{m,k}^{(n)} r_{m,k}^{(n)}(\mathbf{p}^{(n)}) \tag{5.4a}$$

$$\text{s.t.} \sum_{k \in \mathscr{K}_0} \sum_{n \in \mathscr{N}} \rho_{0,k}^{(n)} r_{0,k}^{(n)}(\mathbf{p}^{(n)}) \geq R_{\min}, \tag{5.4b}$$

$$\sum_{k \in \mathscr{K}_m} \rho_{m,k}^{(n)} = 1, \ \forall m \in \mathscr{M}, \ \forall n \in \mathscr{N} \tag{5.4c}$$

$$\rho_{m,k}^{(n)} \in \{0, 1\}, \ \forall n \in \mathscr{N}, \forall m \in \mathscr{M}, \forall k \in \mathscr{K}_m \tag{5.4d}$$

$$\sum_{n \in \mathscr{N}} p_m^{(n)} \leq P_m^{\max}, \ \forall m \in \mathscr{M} \tag{5.4e}$$

$$0 \leq p_m^{(n)} \leq P_m^{(n),\text{mask}}, \ \forall m \in \mathscr{M}, \ \forall n \in \mathscr{N}. \tag{5.4f}$$

In the above formulation, (5.4a) represents the total throughput of all femtocells, whereas (5.4b) imposes the protection of the minimum required macrocell's sum rate $R_{\min}$. Constraints (5.4c) and (5.4d) enforce the OFDMA assumption in each cell m. Transmit powers are constrained by both total power limits given in (5.4e) and spectral masks in (5.4f). To solve (5.4), any direct search method would involve an exhaustive search of all possible subchannel assignment ρ's, followed by finding the optimal power allocation $\mathbf{p}$ for each of these assignments. It is apparent that such an approach incurs an exponential complexity in N—the number of OFDM subchannels, which can be large in practice. Moreover, the non-convexity of the rate function in (5.2) implies that jointly optimizing all transmit powers is in itself a challenging problem, even for a fixed subchannel allotment.

Note that our formulation in (5.4) does not include any instantaneous QoS constraints for individual MUEs and FUEs due to the following reasons. When the channel condition of a certain UE is bad, supporting its instantaneous QoS (expressed in terms of a minimum SINR or data rate) can be infeasible, even if we dedicate a very high transmit power and/or a very large bandwidth to that UE. Furthermore, doing so only creates unnecessary interference to other UEs, while taking away the chances for these UEs to access the limited radio resources. From the system design point of view, it would be sensible to have UEs with unfavorable channel conditions deferring their transmissions and let those with good channels use the resources. This objective can be achieved by maximizing the femtocell total throughput with a macrocell sum rate constraint, as described in (5.4). Because the channel condition of a UE fluctuates over a long period of time, on average each UE can get a fair share of the radio resources, and hence an acceptable average QoS.

Nevertheless, the instantaneous fairness can be provided by imposing individual instantaneous rate constraints. However, it remains unclear whether problem (5.4) with this new set of constraints can be solved, i.e., an open problem for future investigation. Another way to achieve such fairness is to associate each individual rate $r_{m,k}$ ($m \in \mathcal{M}, k \in \mathcal{K}_m$) with a weight $w_{m,k} \geq 0$. Here, a larger value of $w_{m,k}$ means a higher priority given to UE k, and $w_{m,k}$ can be adjusted over time to maintain proportional fairness among the UEs [21]. It is worth noting that a weighted sum-rate constraint, however, makes the subchannel assignment task a challenging one, let alone the power allocation part. While problem (5.4) with a weighted sum-rate constraint for the femtocells is generally difficult and remains unsolved yet, in the specific case of equal weights, i.e., $w_{m,k} = 1$, $\forall m \in \mathcal{M}, k \in \mathcal{K}_m$, such a constraint is actually (5.4b).

5.2   An Iterative Approach to Joint Power and Subchannel Allocation

Towards resolving problem (5.4), we apply the following iterative approach which deals with power allocation and subchannel assignment separately [20, 22, 23]:

$$\underbrace{\boldsymbol{\rho}[0] \rightarrow \mathbf{p}[0]}_{\text{Initialization}} \rightarrow \cdots \rightarrow \underbrace{\boldsymbol{\rho}[t-1] \rightarrow \mathbf{p}[t-1]}_{\text{Iteration } t-1}$$

$$\rightarrow \underbrace{\boldsymbol{\rho}[t] \rightarrow \mathbf{p}[t]}_{\text{Iteration } t} \rightarrow \cdots \rightarrow \underbrace{\boldsymbol{\rho}^{\text{opt}} \rightarrow \mathbf{p}^{\text{opt}}}_{\text{Optimal Solution}}. \tag{5.5}$$

Specifically, we start by computing a feasible solution $(\boldsymbol{\rho}[0], \mathbf{p}[0])$. At the beginning of each iteration t, we find the optimal subchannel assignment $\boldsymbol{\rho}[t]$ for a given power $\mathbf{p}[t-1]$ from the last iteration. Then with the fixed $\boldsymbol{\rho}[t]$, we find the optimal power allocation $\mathbf{p}[t]$. We repeat the process in all subsequent iterations until no further improvement is made. With this iterative approach, we only have to handle two separate subproblems, one at a time—(1) the combinatorial subchannel assignment, and (2) the highly nonconvex power allocation. Since the number of variables is reduced by (almost) half in each optimization subproblem, more tractable solutions to the original problem (5.4) can be developed.

5.2.1   Feasibility and Initial Feasible Allocation

To find a feasible solution $(\rho, \mathbf{p})$ for problem (5.4), the difficulty lies in how one meets the macrocell sum-rate constraint (5.4b). If $R_{\min}$ is greater than the maximum throughput achievable by the macrocell, it is obvious that (5.4) is infeasible. To determine the highest attainable macrocell throughput R^*, we assume that all femtocell BSs do not transmit (i.e., $\mathbf{p}_m = \mathbf{0}$, $\forall m \in \mathcal{M} \backslash \{0\}$), and that each OFDM subchannel is assigned to the MUE with the highest SINR on that subchannel. With these assumptions, the following optimization problem is considered for the macrocell.

$$\max_{\mathbf{p}_0} \sum_{n \in \mathcal{N}} \ln \left(1 + \frac{h_{0,k^*(0,n)}^{(n)} p_0^{(n)}}{\sigma_{k^*(0,n)}^{(n)}}\right) \qquad (5.6)$$

$$\text{s.t.} \sum_{n \in \mathcal{N}} p_0^{(n)} \le P_0^{\max},$$

$$0 \le p_0^{(n)} \le P_0^{(n),\text{mask}}, \ \forall n \in \mathcal{N},$$

where $k^*(0, n) = \arg\max_{k \in \mathcal{K}_0} h_{0,k}^{(n)} p_0^{(n)} / \sigma_k^{(n)}$, $\forall n \in \mathcal{N}$. Since (5.6) is convex, its optimal solution $\mathbf{p}_0^* = [p_0^{(1)*}, p_0^{(2)*}, \dots, p_0^{(N)*}]^T$ can be derived as

$$p_0^{(n)*} = \left[\frac{1}{\alpha} - \frac{\sigma_{k^*(n)}^{(n)}}{h_{0,k^*(n)}^{(n)}}\right]_0^{P_0^{(n),\text{mask}}}, \ \forall n \in \mathcal{N} \qquad (5.7)$$

where $[x]_a^b$ represents the Euclidean projection of x on $[a, b]$, and $\alpha > 0$ is chosen such that $\sum_{n \in \mathcal{N}} p_0^{(n)*} = P_0^{\max}$. Note that the corresponding optimal value $R^* = \sum_{n \in \mathcal{N}} r_{0,k^*(n)}^{(n)}(p_0^{(n)*})$ is the maximum sum rate achievable by all MUEs.

Proposition 5.1. *When the optimal solution of (5.4) is found, (5.4b) is met with equality.*

Proof. The proof can be found in [12].   □

With a fixed $R_{\min}$, the snapshot model that we consider in this paper allows the instantaneous rate R^* to be used in determining the feasibility of problem (5.4). If $R^* < R_{\min}$, problem (5.4) is infeasible. If $R^* = R_{\min}$, there is no capacity remaining for FUEs to use; and thus, the optimal solution of (5.4) is simply $\mathbf{p}^{\text{opt}} = \text{vec}(\mathbf{p}_0^*, \mathbf{0}_N, \cdots, \mathbf{0}_N)$, which gives R^*. If $R^* > R_{\min}$, we take $\mathbf{p}[0] = \text{vec}(\mathbf{p}_0^*, \mathbf{0}_N, \cdots, \mathbf{0}_N)$ as a feasible initial power vector. In this last case, the initial subchannel assignment $\rho[0]$ is established as follows:

- In the macrocell, each subchannel is assigned to the MUE with the highest channel-to-noise ratio on that subchannel.
- In all femtocells, any subchannel assignment is valid because the transmit powers are zero.

5.2.2 Optimal Subchannel Assignment for Fixed Power Allocation

Given a feasible power $\mathbf{p}[t - 1]$, we attempt to find the optimal subchannel assignment $\rho[t]$ at iteration t. Denote by $k(m, n)$ the UE who is given subchannel $n \in \mathcal{N}$ in cell $m \in \mathcal{M}$. Problem (5.4) is now simplified to:

$$\max_{\rho} \sum_{m \in \mathcal{M} \setminus \{0\}} \sum_{k \in \mathcal{K}_m} \sum_{n \in \mathcal{N}} \rho_{m,k}^{(n)} r_{m,k}^{(n)}(\mathbf{p}^{(n)}[t - 1]) \tag{5.8}$$

$$\text{s.t.} \sum_{k \in \mathcal{K}_0} \sum_{n \in \mathcal{N}} \rho_{0,k}^{(n)} r_{0,k}^{(n)}(\mathbf{p}^{(n)}[t - 1]) \geq R_{\min},$$

$$\sum_{k \in \mathcal{K}_m} \rho_{m,k}^{(n)} = 1, \ \forall m \in \mathcal{M}, \ \forall n \in \mathcal{N}$$

$$\rho_{m,k}^{(n)} \in \{0, 1\}, \forall n \in \mathcal{N}, \forall m \in \mathcal{M}, \forall k \in \mathcal{K}_m.$$

Proposition 5.2. *When the optimal solution of (5.8) is found, each subchannel is assigned to the MUE that offers the highest data rate (i.e., the highest SINR) on that subchannel.*

Proof. The proof can be found in [12]. □

It now remains to find an optimal subchannel assignment for all FUEs. The useful result in Proposition 5.2 helps to avoid an exhaustive search that would normally be required to solve (5.8). To see this, notice that

$$r_{0,k(0,n)^{[t]}}^{(n)}(\mathbf{p}^{(n)}[t - 1]) \geq r_{0,k(0,n)^{[t-1]}}^{(n)}(\mathbf{p}^{(n)}[t - 1]), \tag{5.9}$$

since $k(0, n)^{[t]} = \arg\max_{k \in \mathcal{K}_0} r_{0,k}^{(n)}(\mathbf{p}^{(n)}[t - 1])$. Therefore,

$$\sum_{n \in \mathcal{N}} r_{0,k(0,n)^{[t]}}^{(n)}(\mathbf{p}^{(n)}[t - 1]) \geq \sum_{n \in \mathcal{N}} r_{0,k(0,n)^{[t-1]}}^{(n)}(\mathbf{p}^{(n)}[t - 1]) \geq R_{\min}, \tag{5.10}$$

i.e., the first constraint in (5.8) [or (5.4b)] is already satisfied. The subchannel assignment of FUEs can then be decomposed into M_f problems, each for one femtocell $m \in \mathcal{M} \setminus \{0\}$ as follows:

$$\max_{\rho_{m,k}^{(n)} \in \{0,1\}} \sum_{k \in \mathcal{K}_m} \sum_{n \in \mathcal{N}} \rho_{m,k}^{(n)} r_{m,k}^{(n)}(\mathbf{p}^{(n)}[t - 1]) \tag{5.11}$$

$$\text{s.t.} \sum_{k \in \mathcal{K}_m} \rho_{m,k}^{(n)} = 1, \ \forall n \in \mathcal{N}.$$

Such decomposition is possible because $\mathbf{p}[t-1]$ is fixed and the subchannel assignment $\rho_{m,k}^{(n)}$ of UE $k \in \mathcal{K}_m$ does not affect other BSs $m' \in \mathcal{M} \setminus \{m\}$. The solution of (5.11) is to give each subchannel to the FUE that offers the highest data rate on that subchannel.

In summary, the optimal subchannel assignment for any cell $m \in \mathcal{M}$ at iteration t is that

$$\rho_{m,k}^{(n)}[t] = \rho_{m,k}^{(n)*} = \begin{cases} 1, \text{if } k = \arg \max_{k \in \mathcal{K}_m} r_{m,k}^{(n)}(\mathbf{p}^{(n)}[t-1]) \\ 0, \text{otherwise.} \end{cases} \qquad (5.12)$$

5.2.3 Optimal Power Allocation for Fixed Subchannel Assignment

Suppose that the optimal subchannel assignment $\rho[t]$ has already been determined at iteration t. From (5.4), the problem of finding the optimal power allocation $\mathbf{p}[t]$ for $\rho[t]$ is reduced to:

$$\max_{\mathbf{p}} \sum_{m \in \mathcal{M} \setminus \{0\}} \sum_{n \in \mathcal{N}} r_{m,k(m,n)}^{(n)}(\mathbf{p}^{(n)}) \qquad (5.13)$$

$$\text{s.t.} \sum_{n \in \mathcal{N}} r_{0,k(0,n)}^{(n)}(\mathbf{p}^{(n)}) \geq R_{\min},$$

$$\sum_{n \in \mathcal{N}} p_m^{(n)} \leq P_m^{\max}, \ \forall m \in \mathcal{M}$$

$$0 \leq p_m^{(n)} \leq P_m^{(n),\text{mask}}, \ \forall m \in \mathcal{M}, \ \forall n \in \mathcal{N}.$$

Apparently, problem (5.13) is not convex because the rate function in (5.2) is (highly) nonconcave. To overcome such a major difficulty, we adopt the following SCA approach [13] and find the optimal power allocation in another time scale t_p.

1. Initialize with a feasible point $\mathbf{p}[0]$ and set $t_p := 1$.
2. Form the t_pth convex subproblem (Prob-t_p) by approximating the nonconcave objective function and constraints of (5.13) with some concave function around the previous point $\mathbf{p}[t_p - 1]$.
3. Solve convex subproblem (Prob-t_p) to obtain optimal solution $\mathbf{p}[t_p]$.
4. Update the approximation parameters in Step 2 for the next iteration.
5. Increase $t_p := t_p + 1$, go back to Step 2 and iterate until $\mathbf{p}[t_p]$ converges.

In what follows, we will propose three power optimization solutions, each of which corresponds to a different way of approximation and update in Steps 2 and 4 of the above described SCA approach. In each solution, the convergence of the power allocation process will also be analytically proven.

5.2.3.1   Arithmetic-Geometric Mean Approximation

It can be shown that problem (5.13) is equivalent to:

$$\min_{\mathbf{p}} \prod_{m \in \mathcal{M} \setminus \{0\}} \prod_{n \in \mathcal{N}} \frac{\sum_{j \in \mathcal{M} \setminus \{m\}} h_{j,k(m,n)}^{(n)} p_j^{(n)} + \sigma_{k(m,n)}^{(n)}}{\sum_{j \in \mathcal{M}} h_{j,k(m,n)}^{(n)} p_j^{(n)} + \sigma_{k(m,n)}^{(n)}} \tag{5.14}$$

$$\text{s.t.} \quad \prod_{n \in \mathcal{N}} \frac{\sum_{j \in \mathcal{M} \setminus \{0\}} h_{j,k(0,n)}^{(n)} p_j^{(n)} + \sigma_{k(0,n)}^{(n)}}{\sum_{j \in \mathcal{M}} h_{j,k(0,n)}^{(n)} p_j^{(n)} + \sigma_{k(0,n)}^{(n)}} \le e^{-R_{\min}},$$

$$\sum_{n \in \mathcal{N}} p_m^{(n)} \le P_m^{\max}, \; \forall m \in \mathcal{M}$$

$$0 \le p_m^{(n)} \le P_m^{(n),\text{mask}}, \; \forall m \in \mathcal{M}, \; \forall n \in \mathcal{N}.$$

If we define

$$u_j^{(n)}(p_j^{(n)}) := h_{j,k(m,n)}^{(n)} p_j^{(n)} + \sigma_{k(m,n)}^{(n)}, \tag{5.15}$$

then $y_m^{(n)}(\mathbf{p}^{(n)}) := \sum_{j \in \mathcal{M} \setminus \{m\}} u_j^{(n)}(p_j^{(n)})$ and $v_m^{(n)}(\mathbf{p}^{(n)}) := \sum_{j \in \mathcal{M}} u_j^{(n)}(p_j^{(n)})$ are two posynomials[1]. Because $y_m^{(n)}(\mathbf{p}^{(n)})/v_m^{(n)}(\mathbf{p}^{(n)})$ is a ratio of a posynomial to another posynomial, it is not necessarily a posynomial, leaving problem (5.14) still intractable.

The AGM inequality states that

$$\sum_{j \in \mathcal{M}} \kappa_j^{(n)} u_j^{(n)} \ge \prod_{j \in \mathcal{M}} \left[u_j^{(n)} \right]^{\kappa_j^{(n)}}, \tag{5.16}$$

where $\mathbf{u}^{(n)} = [u_0^{(n)}, u_1^{(n)}, \ldots, u_{M_f}^{(n)}]^T \succ \mathbf{0}, \boldsymbol{\kappa}^{(n)} = [\kappa_0^{(n)}, \kappa_1^{(n)}, \ldots, \kappa_{M_f}^{(n)}]^T \succeq \mathbf{0}$ and $\mathbf{1}^T \boldsymbol{\kappa}^{(n)} = 1$. With (5.16), we approximate the posynomial $v_m^{(n)}(\mathbf{p}^{(n)})$ by a monomial $\underline{v}_m^{(n)}(\mathbf{p}^{(n)})$ as:

$$v_m^{(n)}(\mathbf{p}^{(n)}) \ge \underline{v}_m^{(n)}(\mathbf{p}^{(n)}) = \prod_{j \in \mathcal{M}} \left(\frac{u_j^{(n)}(p_j^{(n)})}{\kappa_j^{(n)}} \right)^{\kappa_j^{(n)}}, \tag{5.17}$$

---

[1] A monomial $\hat{q}(\mathbf{x})$ is defined as $\hat{q}(\mathbf{x}) = c x_1^{\hat{a}_1} x_2^{\hat{a}_2} \ldots x_n^{\hat{a}_n}$, where $c > 0$, $\mathbf{x} = [x_1, x_2, \ldots, x_n]^T \in \mathbb{R}_{++}^n$, and $\hat{\mathbf{a}} = [\hat{a}_1, \hat{a}_2, \ldots, \hat{a}_n]^T \in \mathbb{R}^n$. A posynomial is a (nonnegative) sum of monomials [14].

$\forall m \in \mathcal{M}$, where $\kappa_j^{(n)} = u_j^{(n)}(p_j^{(n)})/v_m^{(n)}(\mathbf{p}^{(n)})$. According to [18, Lemma 1], $\underline{v}_m^{(n)}(\mathbf{p}^{(n)})$ is indeed the best local monomial approximation to $v_m^{(n)}(\mathbf{p}^{(n)})$ near $\mathbf{p}^{(n)}$ in the sense of the first-order Taylor approximation.

With (5.17), $y_m^{(n)}(\mathbf{p}^{(n)})/v_m^{(n)}(\mathbf{p}^{(n)})$ is approximated by $y_m^{(n)}(\mathbf{p}^{(n)})/\underline{v}_m^{(n)}(\mathbf{p}^{(n)})$. The latter is a posynomial because it is the ratio of a posynomial to a monomial. Furthermore, since the product of posynomials is a posynomial, the following approximate subproblem (Prob-t_p) of (5.13) belongs to the class of geometric programs:

$$\min_{\mathbf{p}} \prod_{m \in \mathcal{M} \setminus \{0\}} \prod_{n \in \mathcal{N}} \frac{\sum\limits_{j \in \mathcal{M} \setminus \{m\}} h_{j,k(m,n)}^{(n)} p_j^{(n)} + \sigma_{k(m,n)}^{(n)}}{\underline{v}_m^{(n)}(\mathbf{p}^{(n)})} \tag{5.18}$$

$$\text{s.t.} \prod_{n \in \mathcal{N}} \frac{\sum\limits_{j \in \mathcal{M} \setminus \{0\}} h_{j,k(0,n)}^{(n)} p_j^{(n)} + \sigma_{k(0,n)}^{(n)}}{\underline{v}_m^{(n)}(\mathbf{p}^{(n)})} \leq e^{-R_{\min}},$$

$$\sum_{n \in \mathcal{N}} p_m^{(n)} \leq P_m^{\max}, \ \forall m \in \mathcal{M}$$

$$0 \leq p_m^{(n)} \leq P_m^{(n),\text{mask}}, \ \forall m \in \mathcal{M}, \ \forall n \in \mathcal{N}.$$

A geometric program like (5.18) can easily be transformed into a convex problem via a logarithmic change of variables [14, p. 162]. To update the approximation parameters ready for the next iteration $t_p + 1$ in Step 4 of the SCA approach, the current optimal solution $\mathbf{p}[t_p]$ will be used in (5.17). It is worth noting that as we minimize a lower bound of the objective function (5.13), the accuracy of such an approximation is improved in each iteration upon solving convex subproblem (5.18).

Proposition 5.3. *With AGM approximation (5.17), the SCA approach converges to a locally optimal solution that satisfies the Karush-Kuhn-Tucker (KKT) conditions of the original problem (5.13).*

Proof. The proof can be found in [12]. □

5.2.3.2 Logarithmic Approximation

Instead of directly dealing with the highly nonconcave rate function $r_{m,k(m,n)}^{(n)}(\mathbf{p}^{(n)})$ in (5.13), we resort to the following lower bound [15]:

$$\ln\left(1 + z_m^{(n)}\right) \geq \alpha_m^{(n)} \ln z_m^{(n)} + \beta_m^{(n)}, \tag{5.19}$$

which is tight at $z_m^{(n)} = \bar{z}_m^{(n)} \geq 0$ when

$$\alpha_m^{(n)} = \frac{\bar{z}_m^{(n)}}{1 + \bar{z}_m^{(n)}}, \tag{5.20a}$$

$$\beta_m^{(n)} = \ln(1 + \bar{z}_m^{(n)}) - \frac{\bar{z}_m^{(n)}}{1 + \bar{z}_m^{(n)}} \ln\left(\bar{z}_m^{(n)}\right). \tag{5.20b}$$

Applying approximation (5.19) to $r_{m,k(m,n)}^{(n)}(\mathbf{p}^{(n)}) = \ln\left(1 + \gamma_{m,k(m,n)}^{(n)}(\mathbf{p}^{(n)})\right)$ and utilizing the change of variable $\hat{\mathbf{p}} = \ln(\mathbf{p})$, we arrive at the following approximate subproblem (Prob-t_p):

$$\max_{\hat{\mathbf{p}}} \sum_{m \in \mathcal{M} \backslash \{0\}} \underline{r}_m(e^{\hat{\mathbf{p}}}, \boldsymbol{\alpha}_m, \boldsymbol{\beta}_m) \tag{5.21}$$

$$\text{s.t. } \underline{r}_0(e^{\hat{\mathbf{p}}}, \boldsymbol{\alpha}_0, \boldsymbol{\beta}_0) \geq R_{\min},$$

$$\sum_{n \in \mathcal{N}} e^{\hat{p}_m^{(n)}} \leq P_m^{\max}, \ \forall m \in \mathcal{M},$$

$$0 \leq e^{\hat{p}_m^{(n)}} \leq P_m^{(n),\text{mask}}, \ \forall m \in \mathcal{M}, \ \forall n \in \mathcal{N},$$

where $\underline{r}_m(e^{\hat{\mathbf{p}}}, \boldsymbol{\alpha}_m, \boldsymbol{\beta}_m) := \sum_{n \in \mathcal{N}} \alpha_m^{(n)} \ln\left(\gamma_{m,k(m,n)}^{(n)}(e^{\hat{\mathbf{p}}^{(n)}})\right) + \beta_m^{(n)}$ is a lower bound of $r_{m,k(m,n)}^{(n)}(\mathbf{p}^{(n)})$, and $\boldsymbol{\alpha}_m = [\alpha_m^{(1)}, \alpha_m^{(2)}, \cdots, \alpha_m^{(N)}]^T$, $\boldsymbol{\beta}_m = [\beta_m^{(1)}, \beta_m^{(2)}, \cdots, \beta_m^{(N)}]^T$.

Using the fact that the log-sum-exp function is convex [14, p. 72], it is easy to see that (5.21) is a standard concave maximization problem. In (5.21), note that we only maximize a lower bound of the total femtocell throughput, i.e., of the objective function of (5.13). To eventually solve the original problem (5.13), we tighten the bound in (5.19) by iteratively updating $\alpha_m^{(n)}$ and $\beta_m^{(n)}$ as follows. At iteration $t_p = 0$, we initialize $\alpha_m^{(n)} = 1$ and $\beta_m^{(n)} = 0$. At any subsequent iteration $t_p > 0$, we update them according to (5.20) with $\bar{z}_m^{(n)} = \gamma_{m,k(m,n)}^{(n)}(\mathbf{p}^{(n)}[t_p])$, where $\mathbf{p}^{(n)}[t_p]$ is the optimal solution of (5.21). These updated values of $\alpha_m^{(n)}$ and $\beta_m^{(n)}$ will be used in the next iteration $t_p + 1$, as described in Step 4 of the SCA approach.

Proposition 5.4. *With logarithmic approximation (5.19), the SCA approach generates a sequence of improved feasible solutions, which will finally converge to a locally optimal solution $\mathbf{p}^*$ of (5.13).*

Proof. The proof can be found in [12].   □

5.2.3.3 Difference-of-Two-Concave-Functions (D.C.) Approximation

In this case, we first express the rate function (5.2) in a D.C. form as:

$$\sum_{n \in \mathcal{N}} r_{m,k(m,n)}^{(n)}(\mathbf{p}^{(n)}) := f_m(\mathbf{p}) - g_m(\mathbf{p}), \tag{5.22}$$

where $f_m(\mathbf{p})$ and $g_m(\mathbf{p})$ are the two concave functions defined as follows:

$$f_m(\mathbf{p}) := \sum_{n \in \mathcal{N}} \ln \left(\sum_{j \in \mathcal{M}} h_{j,k(m,n)}^{(n)} p_j^{(n)} + \sigma_{k(m,n)}^{(n)} \right), \tag{5.23}$$

$$g_m(\mathbf{p}) := \sum_{n \in \mathcal{N}} \ln \left(\sum_{j \in \mathcal{M} \setminus \{m\}} h_{j,k(m,n)}^{(n)} p_j^{(n)} + \sigma_{k(m,n)}^{(n)} \right), \tag{5.24}$$

for any BS $m \in \mathcal{M}$.

Then, we employ the following approximation [19]:

$$g_m(\mathbf{p}) \approx g_m(\mathbf{p}[t_p - 1]) + \nabla g_m^T(\mathbf{p}[t_p - 1]) \left(\mathbf{p} - \mathbf{p}[t_p - 1] \right) \tag{5.25}$$

for a fixed $\mathbf{p}[t_p - 1]$ from iteration $t_p - 1 \geq 0$. Here, $\nabla g_m(\mathbf{p})$ is a vector of length $(M_f + 1)N$ with its entry defined as

$$\nabla g_m(\mathbf{p})^{(Nj+n)} := \begin{cases} 0, & \text{if } j = m, \\ \dfrac{h_{j,k(m,n)}^{(n)}}{\displaystyle\sum_{s \in \mathcal{M} \setminus \{m\}} h_{s,k(m,n)}^{(n)} p_s^{(n)} + \sigma_{k(m,n)}^{(n)}}, & \text{if } j \in \mathcal{M} \setminus \{m\}, \end{cases} \tag{5.26}$$

for $n \in \mathcal{N}$. From (5.22)–(5.25), it is clear that

$$\sum_{n \in \mathcal{N}} r_{m,k(m,n)}^{(n)}(\mathbf{p}^{(n)}) \approx f_m(\mathbf{p}) - g_m(\mathbf{p}[t_p - 1]) \tag{5.27}$$

$$- \nabla g_m^T(\mathbf{p}[t_p - 1]) \left(\mathbf{p} - \mathbf{p}[t_p - 1] \right),$$

the right-hand side of which is actually a concave function with respect to $\mathbf{p}$.

The approximation in (5.27) allows us to recast (5.13) into a sequence of convex optimization subproblems as follows [17]. Starting from a feasible $\mathbf{p}[0]$, the optimal solution $\mathbf{p}[t_p]$ at iteration $t_p > 0$ is determined upon solving the following convex program (Prob-t_p):

$$\max_{\mathbf{p}} \sum_{m \in \mathcal{M} \setminus \{0\}} f_m(\mathbf{p}) - g_m(\mathbf{p}[t_p - 1]) - \nabla g_m^T(\mathbf{p}[t_p - 1]) \left(\mathbf{p} - \mathbf{p}[t_p - 1]\right) \quad (5.28)$$

$$\text{s.t.} \quad f_0(\mathbf{p}) - g_0(\mathbf{p}[t_p - 1]) - \nabla g_0^T(\mathbf{p}[t_p - 1]) \left(\mathbf{p} - \mathbf{p}[t_p - 1]\right) \geq R_{\min},$$

$$\sum_{n \in \mathcal{N}} p_m^{(n)} \leq P_m^{\max}, \ \forall m \in \mathcal{M}$$

$$0 \leq p_m^{(n)} \leq P_m^{(n),\text{mask}}, \ \forall m \in \mathcal{M}, \ \forall n \in \mathcal{N}.$$

where $\mathbf{p}[t_p - 1]$ has already been found from the last iteration $t_p - 1$. The value of $\mathbf{p}[t_p]$ will then be used in the approximation (5.25) to find the optimal solution of problem (Prob-$(t_p + 1)$) in the next iteration $t_p + 1$.

Proposition 5.5. *With D.C. approximation (5.27), the SCA approach generates a sequence of improved feasible solutions, which will finally converge to a locally optimal solution $\mathbf{p}^*$ of (5.13).*

Proof. The proof can be found in [12]. □

5.3 Proposed Joint Power and Subchannel Allocation Algorithms with Macrocell Total Throughput Protection

We summarize in Algorithm 5.1 the proposed iterative algorithm that jointly allocates subchannels and powers in an OFDMA-based mixed macrocell/femtocell network. Specifically, the algorithm starts with finding a feasible allocation, using the result derived in Sect. 5.2.1. Given a fixed power allocation $\mathbf{p}[t - 1]$, the optimal subchannel assignment $\rho[t]$ at iteration $t > 0$ is determined upon applying the solution in Sect. 5.2.2. Then, for a fixed $\rho[t]$, the optimal power allocation $\mathbf{p}[t]$ is found by any of the three SCA approach-based schemes described in Sect. 5.2.3 (i.e., AGM, logarithmic, and D.C. approximations). The process repeats until $\mathbf{p}$ and ρ converge.

Algorithm 5.1 Proposed Iterative Subchannel and Power Allocation

---
1: Initialize: $t := 1$
2: Compute $\mathbf{p}[0] = \text{vec}(\mathbf{p}_0^*, \mathbf{0}_N, \cdots, \mathbf{0}_N)$ according to (5.7).
3: **repeat** {To solve (5.4)}
4:      For a fixed $\mathbf{p}[t - 1]$, find optimal subchannel assignment $\rho[t]$ using (5.12).
5:      For a fixed $\rho[t]$, find optimal power allocation $\mathbf{p}[t]$ by solving (5.13) with the SCA approach, i.e., solving a series of subproblems (Prob-t_p) in (5.18) [by AGM approximation], or in (5.21) [by logarithmic approximation], or in (5.28) [by D.C. approximation].
6:      Set $t := t + 1$.
7: **until** Convergence of $\mathbf{p}$ and ρ

Proposition 5.6. *For a feasible problem (5.4), Algorithm 5.1 will converge to give a local maximum of (5.4).*

Proof. The proof can be found in [12].   □

Note that for a fixed subchannel assignment $\rho[t]$, the performance gap from the locally optimal $\mathbf{p}[t]$ found by the proposed SCA approach to the actual globally optimal $\mathbf{p}^*[t]$ is unknown. Because these two power allocation solutions can be different, their corresponding subchannel assignments $\rho[t + 1]$ and $\rho^*[t + 1]$ in the next iteration $t + 1$ can also be different; and so are their corresponding final joint subchannel and power allocation solutions. While one may believe that the iterative method (5.5) with a globally optimal power allocation would outperform that with any local power optimization, the proof of such is unavailable. What can be proven is that these joint solutions both give some local maxima of (5.4). Nevertheless, it should be mentioned that the SCA approach often empirically achieves the globally optimal power allocation [15, 18, 19].

In Algorithm 5.1, it is important to note that the initial allocation (5.7) can be performed at the macrocell BS. As well, the subchannel assignment (5.12) can be locally executed at each BS $m \in \mathcal{M}$. Algorithm 5.1 can thus be implemented centrally or distributively, depending on how the SCA-based solution to the power allocation problem (5.13) [and, in turn, subproblem (Prob-t_p) in (5.18), (5.21), and (5.28)] is realized.

5.3.1 Centralized SCA-based Power Allocation with AGM Approximation

From the discussion in Sect. 5.2.3.1, we present in Algorithm 5.2 a power allocation scheme based on the SCA approach with the AGM approximation to solve (5.13). In Algorithm 5.2, there are two key steps: (1) Compute the lower-bound monomial $\underline{v}_m^{(n)[t_p]}(\mathbf{p}^{(n)})$ [see (5.17)] at iteration t_p, and (2) Solve a geometric (i.e. convex) program [see subproblem (Prob-t_p) in (5.18)] by an interior-point method. To complete these steps, one must know all the terms $u_j^{(n)}(p_j^{(n)}[t_p - 1])$ and $\kappa_j^{(n)}[t_p]$ in $\underline{v}_m^{(n)[t_p]}(\mathbf{p}^{(n)})$ and rely on a centralized convex solver. To implement Algorithm 5.2, a central processing unit is most likely needed to collect all the network information (e.g., $h_{j,k(m,n)}^{(n)}$ and $\sigma_{k(m,n)}^{(n)}$) and perform the power optimization.

Placed at an OAM server in the core network, such a processing unit is capable of collecting information about both macrocell and femtocells, executing the optimization task, and disseminating the computed solutions to all BSs under its management. The exchange of network information and control signals can be made possible via the residential broadband access links, e.g., DSL, to which all femtocell BSs are connected.

Algorithm 5.2 Centralized SCA-based Power Allocation with Arithmetic-Geometric Mean Approximation

---

1: Initialize: $t_p := 1$.
2: **repeat** {To solve (5.13)}
3:     Compute each term $u_j^{(n)}(p_j^{(n)}[t_p - 1])$, $\forall j \in \mathcal{M}, \forall n \in \mathcal{N}$ in (5.15).
4:     Compute each coefficient

$$\kappa_j^{(n)}[t_p] = \frac{u_j^{(n)}(p_j^{(n)}[t_p - 1])}{\displaystyle\sum_{j \in \mathcal{M}} u_j^{(n)}(p_j^{(n)}[t_p - 1])}, \quad \forall j \in \mathcal{M}, \forall n \in \mathcal{N}.$$

5:     Compute monomial

$$\underline{v}_m^{(n)[t_p]}(\mathbf{p}^{(n)}) = \prod_{j \in \mathcal{M}} \left(\frac{u_j^{(n)}(p_j^{(n)}[t_p - 1])}{\kappa_j^{(n)}[t_p]} \right)^{\kappa_j^{(n)}[t_p]}.$$

6:     With $\underline{v}_m^{(n)[t_p]}(\mathbf{p}^{(n)})$, solve (convex) geometric program (5.18), e.g., by an interior-point method, for an optimal power $\mathbf{p}[t_p]$.
7:     Set $t_p := t_p + 1$.
8: **until** Convergence of $\mathbf{p}$

5.3.2 Distributed SCA-based Power Allocation with Logarithmic Approximation

In femtocell networks with limited backhaul capacity and without a central processing unit (e.g., when macrocell and femtocells belong to different service providers), it can be more desirable to distributively implement the resource allocation solutions. Utilizing Lagrangian duality, we will show that the SCA-based power allocation based on logarithmic and D.C. approximations can be performed by individual BSs with limited information exchange.

Let $\phi \geq 0$ and $\boldsymbol{\varphi} = [\varphi_m]_{m \in \mathcal{M}} \succeq \mathbf{0}$ be the Lagrangian multipliers of subproblem (Prob-t_p) in (5.21). The Lagrangian of (5.21) is defined as:

$$\mathcal{L}(\hat{\mathbf{p}}, \phi, \boldsymbol{\varphi}) := \sum_{m \in \mathcal{M} \setminus \{0\}} \underline{r}_m(e^{\hat{\mathbf{p}}}, \alpha_m, \beta_m) + \phi \left(\underline{r}_0(e^{\hat{\mathbf{p}}}, \alpha_0, \beta_0) - R_{\min} \right)$$

$$- \sum_{m \in \mathcal{M}} \varphi_m \left(\sum_{n \in \mathcal{N}} e^{\hat{p}_m^{(n)}} - P_m^{\max} \right), \tag{5.29}$$

where $\alpha_m, \beta_m, \forall m \in \mathcal{M}$ are fixed constants. The dual function of which is then given by:

$$D(\phi, \boldsymbol{\varphi}) := \max_{\hat{\mathbf{p}}} \mathcal{L}(\hat{\mathbf{p}}, \phi, \boldsymbol{\varphi}). \tag{5.30}$$

Upon solving the stationary condition $\partial \mathscr{L}(\hat{\mathbf{p}}, \phi, \varphi)/\partial \hat{p}_m^{(n)} = 0$ [24] and transforming the result back to the **p**-space, the following fixed-point equation can be derived which maximizes (5.30):

$$p_m^{(n)} = \left[\frac{\bar{\phi}_m \alpha_m^{(n)}}{\displaystyle\sum_{s \in \mathscr{M} \backslash \{m\}} \frac{\bar{\phi}_s \alpha_s^{(n)}}{\mathscr{I}_{k(s,n)}^{(n)}(\mathbf{p}_{-s}^{(n)})} h_{m,k(s,n)}^{(n)} + \varphi_m} \right]_0^{P_m^{(n),\text{mask}}} \tag{5.31}$$

where we define

$$\bar{\phi}_{\bar{s}} := \begin{cases} \phi, & \text{if } \bar{s} = 0, \\ 1, & \text{if } \bar{s} \in \mathscr{M} \backslash \{0\}, \end{cases} \tag{5.32}$$

$$\mathscr{I}_{k(s,n)}^{(n)}(\mathbf{p}_{-s}^{(n)}) := \sum_{s' \in \mathscr{M} \backslash \{s\}} h_{s',k(s,n)}^{(n)} p_{s'}^{(n)} + \sigma_{k(s,n)}^{(n)}. \tag{5.33}$$

Once an optimal power allocation is found by (5.31), solution of the dual problem $\min_{\phi > 0, \varphi \succ 0} D(\phi, \varphi)$ can be determined by a subgradient method as follows.

$$\phi[t_s + 1] = \left[\phi[t_s] + \delta_\phi \left(R_{\min} - \underline{r}_0(\mathbf{p}, \alpha_0, \beta_0) \right) \right]^+, \tag{5.34}$$

$$\varphi_m[t_s + 1] = \left[\varphi_m[t_s] + \delta_\varphi \left(\sum_{n \in \mathscr{N}} p_m^{(n)} - P_m^{\max} \right) \right]^+, \tag{5.35}$$

$\forall m \in \mathscr{M}$, where $[\cdot]^+ = \max(\cdot, 0)$ and $\delta_\phi, \delta_\varphi > 0$ are step sizes.

Combining the preceding derivations with the results in Sect. 5.2.3.2, we present in Algorithm 5.3 a distributed SCA-based power allocation scheme to solve (5.13). Specifically, the inner loop is to compute an optimal power management policy for a given value of α and β, i.e., to resolve subproblem (Prob-t_p) in (5.21). Note that it is not necessary to fully optimize $p_m^{(n)}$ before updating Lagrangian multipliers ϕ and φ. In practice, we only need a single iteration of (5.31). On the other hand, the outer loop is to update α and β; and thus, tightening the bound in (5.19). As can be seen, each step of Algorithm 5.3 can be executed by an individual BS $m \in \mathscr{M}$ with:

1. *Interference* terms $\mathscr{I}_{k(m,n)}^{(n)}(\mathbf{p}_{-m}^{(n)}[t_s])$ locally measured and fed back by its own UEs $k(m,n) \in \mathscr{K}_m$.
2. *Scaled inverse interference* terms $\mathscr{I}_{k(s,n)}^{(n)}(\mathbf{p}_{-s}^{(n)}[t_s])$ broadcast by other BSs $s \in \mathscr{M} \backslash \{m\}$, where we define

---

Algorithm 5.3 Distributed SCA-based Power Allocation with Logarithmic Approximation

---

1: Initialize: $t_p := 0$ and $t_s := 0$.
2: Macrocell BS initializes $\phi[0] > 0$.
3: BS $m \in \mathcal{M}$ initializes $\varphi_m[0] > 0$, $p_m^{(n)}[0] := 0$, $\alpha_m^{(n)}[0] := 1$, $\beta_m^{(n)}[0] := 0$, $\forall m \in \mathcal{M}, \forall n \in \mathcal{N}$.
4: **repeat** {To solve (5.13)}
5:    **repeat** {To solve (5.21)}
6:       UE $k(s,n) \in \mathcal{K}_s$ measures and reports $\mathscr{I}_{k(s,n)}^{(n)}(\mathbf{p}_{-s}^{(n)}[t_s])$ [see (5.33)] to its BS $s \in \mathcal{M}$.
7:       BS $s \in \mathcal{M}$ computes and broadcasts $\mathscr{J}_{k(s,n)}^{(n)}(\mathbf{p}_{-s}^{(n)}[t_s])$ [see (5.36)] to other BSs $s' \in \mathcal{M} \backslash \{s\}$.
8:       BS $m \in \mathcal{M}$ computes its power $p_m^{(n)}$, $\forall n \in \mathcal{N}$ by:

$$p_m^{(n)}[t_s + 1] = \left[\frac{\bar{\phi}_m[t_s]\alpha_m^{(n)}[t_p]}{\sum\limits_{s \in \mathcal{M} \backslash \{m\}} \mathscr{J}_{k(s,n)}^{(n)}(\mathbf{p}_{-s}^{(n)}[t_s])h_{m,k(s,n)}^{(n)} + \varphi_m[t_s]} \right]_0^{P_m^{(n),\text{mask}}}$$

9:       Macrocell BS updates ϕ by (5.34).
10:      BS $m \in \mathcal{M}$ updates φ_m by (5.35).
11:      Set $t_s := t_s + 1$.
12:   **until** ϕ and φ converge
13:   Set $\mathbf{p}^*[t_p] := \mathbf{p}[t_s]$.
14:   BS $m \in \mathcal{M}$ updates $\alpha_m^{(n)}[t_p + 1]$ and $\beta_m^{(n)}[t_p + 1]$ using (5.20) with $\bar{z}_m^{(n)} = \gamma_{m,k(m,n)}^{(n)}(\mathbf{p}^{*(n)}[t_p])$.
15:   Set $t_p := t_p + 1$.
16: **until** p converges

$$\mathscr{J}_{k(s,n)}^{(n)}(\mathbf{p}_{-s}^{(n)}[t_s]) := \frac{\bar{\phi}_s[t_s]\alpha_s^{(n)}[t_p]}{\mathscr{I}_{k(s,n)}^{(n)}(\mathbf{p}_{-s}^{(n)}[t_s])}. \tag{5.36}$$

3. *Channel gains* $h_{m,k(s,n)}^{(n)}$ measured and fed back by UEs $k(s,n) \in \mathcal{K}_s, s \in \mathcal{M} \backslash \{m\}$. Since we assume a block fading model, channel information only needs to be sent once at the beginning of the optimization process [8, 25, 26].

For fixed multipliers ϕ and φ, it is shown in [15, Lemma 3] that the fixed-point power update (5.31) always converges to the maximizer of $\mathcal{L}(\hat{\mathbf{p}}, \phi, \varphi)$ in (5.29). Using small values of $\delta_\phi, \delta_\varphi$, the updates of ϕ and φ in (5.34)–(5.35) are also guaranteed to converge [24]. Together with the result in Proposition 5.4, it can be concluded that the decentralized Algorithm 5.3 converges to an optimal solution of problem (5.13).

5.3.3 Distributed SCA-based Power Allocation with D.C. Approximation

In this case, let $\lambda \geq 0$ and $\boldsymbol{\mu} = [\mu_m]_{m \in \mathcal{M}} \succeq \mathbf{0}$ be the Lagrangian multipliers. The Lagrangian of subproblem (Prob-t_p) in (5.28) is now:

$$
\begin{aligned}
\mathscr{L}(\mathbf{p}, \lambda, \boldsymbol{\mu}) := & \sum_{m \in \mathcal{M}} \bar{\lambda} \sum_{n \in \mathcal{N}} \ln \left(\sum_{s \in \mathcal{M}} h_{s,k(m,n)}^{(n)} p_s^{(n)} + \sigma_{k(m,n)}^{(n)} \right) \\
& - \sum_{m \in \mathcal{M}} \bar{\lambda} \sum_{n \in \mathcal{N}} \left(\sum_{s \in \mathcal{M} \backslash \{m\}} a_{s,m}^{(n)} p_s^{(n)} + c_m^{(n)} \right) \\
& - \lambda R_{\min} - \sum_{m \in \mathcal{M}} \mu_m \left(\sum_{n \in \mathcal{N}} p_m^{(n)} - P_m^{\max} \right),
\end{aligned}
\tag{5.37}
$$

where we define

$$
\bar{\lambda} := \begin{cases} \lambda, & \text{if } m = 0, \\ 1, & \text{if } m \in \mathcal{M} \backslash \{0\}, \end{cases}
\tag{5.38}
$$

$$
a_{s,m}^{(n)} := \frac{h_{s,k(m,n)}^{(n)}}{\mathscr{I}_{k(m,n)}^{(n)}(\mathbf{p}_{-s}^{(n)}[t_p - 1])},
\tag{5.39}
$$

$$
c_m^{(n)} := \ln \left(\mathscr{I}_{k(m,n)}^{(n)}(\mathbf{p}_{-s}^{(n)}[t_p - 1]) \right) + \frac{\sigma_{k(m,n)}^{(n)}}{\mathscr{I}_{k(m,n)}^{(n)}(\mathbf{p}_{-s}^{(n)}[t_p - 1])} - 1.
\tag{5.40}
$$

Note here that $a_{s,m}^{(n)}$ and $c_m^{(n)}$ are fixed in the current iteration t_p because they are based on the value of $\mathbf{p}_{-s}^{(n)}[t_p - 1]$, which has already been found in the last iteration $t_p - 1$. The dual function of (5.28) can then be written as:

$$
D(\lambda, \boldsymbol{\mu}) := \max_{\mathbf{p}} \mathscr{L}(\mathbf{p}, \lambda, \boldsymbol{\mu}).
\tag{5.41}
$$

From the stationary condition $\partial \mathscr{L}(\mathbf{p}, \lambda, \boldsymbol{\mu}) / \partial p_m^{(n)} = 0$, the following fixed-point equation maximizes (5.41):

$$
p_m^{(n)} = \left[\frac{\bar{\lambda}}{\mu_m + \bar{\lambda} \sum_{s \in \mathcal{M} \backslash \{m\}} h_{m,k(s,n)}^{(n)} \left(\dfrac{1}{\mathscr{I}_{k(s,n)}^{(n)}(\mathbf{p}_{-s}^{(n)}[t_p - 1])} - \dfrac{1}{\mathscr{I}_{k(s,n)}^{(n)}(\mathbf{p}_{-s}^{(n)}) + h_{s,k(s,n)}^{(n)} p_s^{(n)}} \right)} - \frac{\mathscr{I}_{k(m,n)}^{(n)}(\mathbf{p}_{-m}^{(n)})}{h_{m,k(m,n)}^{(n)}} \right]_0^{P_m^{(n),\text{mask}}}
\tag{5.43}
$$

With an optimal power allocation found by (5.43), solution of the dual problem $\min_{\lambda > 0, \mu \succ 0} D(\lambda, \mu)$ can be determined by a subgradient method as follows.

$$\lambda[t_s + 1] = \left[\lambda[t_s] + \delta_\lambda \left\{ \sum_{n \in \mathcal{N}} \frac{\mathcal{I}_{k(0,n)}^{(n)}(\mathbf{p}_{-0}^{(n)})}{\mathcal{I}_{k(0,n)}^{(n)}(\mathbf{p}_{-0}^{(n)}[t_p - 1])} - N \right. \right.$$

$$\left. \left. + \ln \left(\frac{\mathcal{I}_{k(0,n)}^{(n)}(\mathbf{p}_{-0}^{(n)}[t_p - 1])}{\mathcal{I}_{k(0,n)}^{(n)}(\mathbf{p}_{-0}^{(n)}) + h_{0,k(0,n)}^{(n)} p_0^{(n)}} \right) + R_{\min} \right\} \right]^+ , \qquad (5.44)$$

$$\mu_m[t_s + 1] = \left[\mu_m[t_s] + \delta_\mu \left(\sum_{n \in \mathcal{N}} p_m^{(n)} - P_m^{\max} \right) \right]^+ , \qquad (5.45)$$

$\forall m \in \mathcal{M}$, where $\delta_\lambda, \delta_\mu > 0$ are step sizes.

From the above derivations and the results presented in Sect. 5.2.3.3, we propose in Algorithm 5.4 a distributed SCA-based power allocation scheme with the following power update:

$$\tilde{p}_m^{(n)}[t_s + 1] = \left[\frac{\bar{\lambda}[t_s]}{\mu_m[t_s] + \bar{\lambda}[t_s] \sum_{s \in \mathcal{M} \backslash \{m\}} h_{m,k(s,n)}^{(n)} \left(\frac{1}{\mathcal{I}_{k(s,n)}^{(n)}(\mathbf{p}_{-s}^{(n)}[t_p - 1])} - \frac{1}{\mathcal{Q}_{k(s,n)}^{(n)}(\tilde{\mathbf{p}}^{(n)}[t_s])} \right)} \right.$$

$$\left. - \frac{\mathcal{I}_{k(m,n)}^{(n)}(\tilde{\mathbf{p}}_{-m}^{(n)}[t_s])}{h_{m,k(m,n)}^{(n)}} \right]_0^{P_m^{(n),\mathrm{mask}}} . \qquad (5.47)$$

Specifically, the inner loop is to compute an optimal power solution for subproblem (Prob-t_p) in (5.28). The outer loop is to update the D.C. approximation in (5.25) for the next iteration, i.e., to finally solve the power allocation problem (5.13). Rather similar to Algorithm 5.3, each step of Algorithm 5.4 can be executed by an individual BS $m \in \mathcal{M}$ with:

1. *Interference* terms $\mathcal{I}_{k(m,n)}^{(n)}(\mathbf{p}_{-m}^{(n)}[t_s])$ locally measured and fed back by its own UEs $k(m,n) \in \mathcal{K}_m$.
2. *Interference* terms $\mathcal{I}_{k(s,n)}^{(n)}(\mathbf{p}_{-s}^{(n)}[t_s])$ and *aggregate interference* terms $\mathcal{Q}_{k(s,n)}^{(n)}$ broadcast by other BSs $s \in \mathcal{M} \backslash \{m\}$, where we define

$$\mathcal{Q}_{k(s,n)}^{(n)}(\tilde{\mathbf{p}}^{(n)}[t_s]) := \mathcal{I}_{k(s,n)}^{(n)}(\tilde{\mathbf{p}}_{-s}^{(n)}[t_s]) + h_{s,k(s,n)}^{(n)} \tilde{p}_s^{(n)}[t_s]. \qquad (5.48)$$

3. *Channel gains* $h_{m,k(s,n)}^{(n)}$ measured and fed back by UEs $k(s,n) \in \mathcal{K}_s, s \in \mathcal{M} \backslash \{m\}$.

For given multipliers λ and μ, it is shown in [27] that the fixed-point power update (5.43) converges to the maximizer of $\mathcal{L}(\mathbf{p}, \lambda, \mu)$ in (5.37) under some mild conditions. With small $\delta_\lambda, \delta_\mu$, the updates of λ and μ in (5.44)–(5.45) are also

---

Algorithm 5.4 Distributed SCA-based Power Allocation with D.C. Approximation

---

1: Initialize: $t_p := 1$ and $t_s := 0$.
2: Macrocell BS initializes $\lambda[0] > 0$.
3: BS $m \in \mathcal{M}$ initializes $\mu_m[0] > 0$ and $\tilde{p}_m^{(n)}[0] = 0$, $\forall m \in \mathcal{M}$, $\forall n \in \mathcal{N}$.
4: **repeat** {To solve (5.13) with optimizing variable $\mathbf{p}$}
5:     **repeat** {To solve (5.28) with optimizing variable $\tilde{\mathbf{p}}$}
6:         UE $k(s,n) \in \mathcal{K}_s$ measures and reports $h_{s,k(s,n)}^{(n)}$ and $\mathcal{I}_{k(s,n)}^{(n)}(\tilde{\mathbf{p}}_{-s}^{(n)}[t_s])$ [see (5.33)] to its BS
        $s \in \mathcal{M}$.
7:         BS $s \in \mathcal{M}$ computes and broadcasts $\mathcal{Q}_{k(s,n)}^{(n)}(\tilde{\mathbf{p}}^{(n)}[t_s])$ [see (5.48)] to other BSs $s' \in$
        $\mathcal{M}\backslash\{s\}$.
8:         BS $m \in \mathcal{M}$ computes its power $\tilde{p}_m^{(n)}$ by (5.47) for all $n \in \mathcal{N}$.
9:         Macrocell BS updates λ by (5.44) (with $\mathbf{p}_{-0}^{(n)}$ replaced by $\tilde{\mathbf{p}}_{-0}^{(n)}$).
10:         BS $m \in \mathcal{M}$ updates μ_m by (5.45) (with $p_m^{(n)}$ replaced by $\tilde{p}_m^{(n)}$).
11:         Set $t_s := t_s + 1$.
12:     **until** λ and μ converge
13:     Set $\mathbf{p}[t_p] := \tilde{\mathbf{p}}[t_s]$.
14:     BS $s \in \mathcal{M}$ broadcasts $\mathcal{I}_{k(s,n)}^{(n)}(\mathbf{p}_{-s}^{(n)}[t_p])$ to other BSs $s' \in \mathcal{M}\backslash\{s\}$.
15:     Set $t_p := t_p + 1$.
16: **until** $\mathbf{p}$ converges

---

guaranteed to converge, resulting in an optimal solution for (5.28) [24]. Together with the result in Proposition 5.5, the decentralized Algorithm 5.4 will converge to an optimal solution of problem (5.13).

5.4 Illustrative Results

In our numerical examples we reuse the network topology shown in Fig. 4.3, where MUEs and FUEs are randomly deployed inside circles of radii of 500 m and 50 m, respectively. Assume there are $M = 10$ MUEs in the macrocell while $K = 20$ FUEs are equally divided among four femtocells. In each cell, we assume OFDMA downlink transmissions with $N = 8$ subchannels, each of which has a total bandwidth of 180 KHz. At the macrocell BS, we set $P_0^{\max} = 47$ dBm whereas at each femtocell BS $P_m^{\max} = 23$ dBm, $\forall m \in \mathcal{M}\backslash\{0\}$. The spectral mask is chosen as $P_m^{\mathrm{mask}} = P_m^{\max}/N$, $\forall m \in \mathcal{M}$, and noise power density -174 dBm/Hz. Channel gains are set as $h_{m,k}^{(n)} = \chi^{(n)}d_{m,k}^{-\beta}$, where $\chi^{(n)}$ is a random value generated according to the Rayleigh distribution, $d_{m,k}$ is the geographical distance between BS m and UE k, and $\beta = 3$ is the pathloss exponent.

We present in Fig. 5.2 the overall convergence process of the proposed Algorithm 5.1, which involves both subchannel assignment and power allocation. Clearly, as the average femtocell throughput improves after every iteration, the algorithm eventually converges after some tens of iterations. In Fig. 5.2a, it is observed that the remaining capacity available for FUEs is actually reduced for larger values of R^{min}. On the other hand, Fig. 5.2b indicates that the total macrocell throughput is always

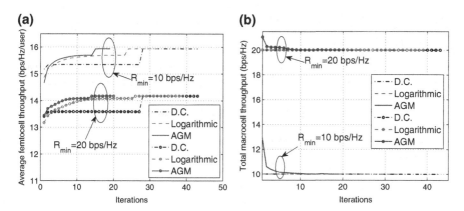

Fig. 5.2 Performance of Algorithm 5.1 (both power and subchannel allocation. (**a**) Femtocell throughput. (**b**) Macrocell throughput)

above the specified threshold $R^{\min}$. Not only is such a numerical result consistent with our finding in Proposition 5.1, it also confirms that the proposed Algorithm 5.1 is capable of protecting the network capacity of the macrocell at all times.

Also can be seen from Fig. 5.2a is that AGM approximation in the SCA-based power allocation (see Algorithm 5.2) gives Algorithm 5.1 the shortest convergence time. This is most likely due to the centralized nature of Algorithm 5.2, where the power optimization is accomplished in one shot. For a given subchannel assignment, Fig. 5.3 shows that the distributed SCA power allocation schemes based on the logarithmic approximation (see Algorithm 5.3) and D.C. approximation (see Algorithm 5.4) quickly converge after fewer than ten iterations.

Figure 5.4 illustrates the convergence behavior of Algorithm 5.1 with different initial allocations. For simplicity, we assume that $R^{\min} = 0$. In this case, any initial power allocation is feasible as long as it satisfies the power constraints in (5.4e) and (5.4f). To initialize the subchannel assignment, we give each subchannel to the UE with the highest SINR on that subchannel. Our numerical results confirm that the SCA-based power allocations given by AGM, logarithmic and D.C. approximations (see Algorithms 5.2–5.4, respectively) always converge for all the chosen values of $\mathbf{p}_{\text{initial}}$. Again, Fig. 5.4 shows that the joint power and subchannel allocation based on AGM approximation gives the shortest convergence time, whereas that based on the other two approximations exhibit quite similar convergence behaviors. Although starting at different values of average femtocell throughput, Algorithm 5.1 used in combination with these three approximations converges to the same final optimal value.

With $R^{\min} = 0$ and $\mathbf{p}_{\text{initial}} = \mathbf{P}^{\text{mask}} = \mathbf{P}^{\text{max}}/N$, Fig. 5.5 displays the assignment of subchannels by the proposed Algorithm 5.1. As can be seen, all three approximations eventually result in an identical allocation, except for subchannel 7 in AGM approximation.Notably, some UEs are assigned with most of the subchannels in

Fig. 5.3 Convergence of
Algorithms 5.3 and 5.4
(distributed power allocation)

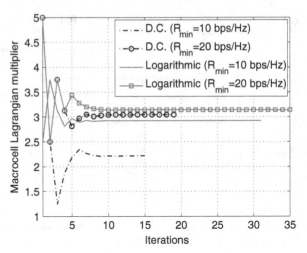

Fig. 5.4 Convergence of
Algorithm 5.1 with different
initial allocations ($R^{min} = 0$)

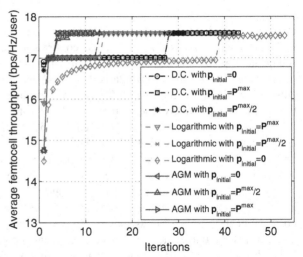

their own cell (e.g., MUE 0 and FUE 26). One possible explanation is that the channel conditions of these UEs are highly favorable in the realization that we randomly generate.

Finally, Fig. 5.6 demonstrates the tradeoff between the per-FUE throughput and macrocell's R^{min}. It is clear from the figure that all three SCA-based power allocation schemes give Algorithm 5.1 very similar performance. Also note that by reducing the minimum total network capacity required by the macrocell from 20 to 0 bps/Hz, the improvement in femtocell's sum throughput far exceeds threefold the amount of rate loss in the macrocell. Such a significant gain can be explained by realizing that the FUEs are located very close to their corresponding BSs, and thus able to achieve much higher data rates compared to those by the MUEs.

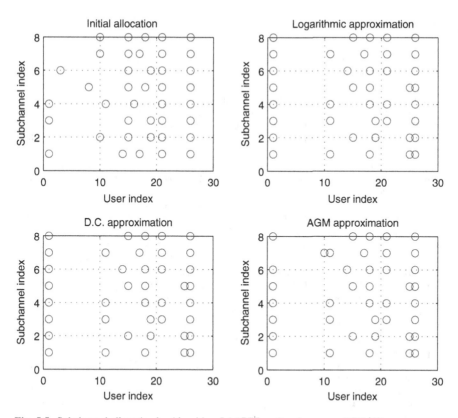

Fig. 5.5 Subchannel allocation by Algorithm 5.1 ($R^{\min} = 0$ and $\mathbf{p}_{\text{initial}} = \mathbf{P}^{\max}/N$)

Fig. 5.6 Tradeoff between femtocell throughput and macrocell $R^{\min}$

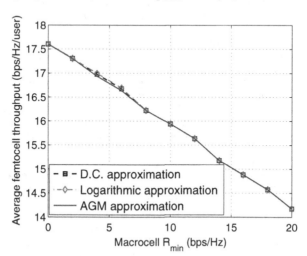

References

1. G. d. l. Roche, A. Valcarce, D. Lopez-Perez, and J. Zhang, "Access control mechanisms for femtocells," *IEEE Commun. Mag.*, vol. 48, no. 1, pp. 33–39, Jan. 2010.
2. M. Yavuz, F. Meshkati, S. Nanda, A. Pokhariyal, N. Johnson, B. Raghothaman, and A. Richardson, "Interference management and performance analysis of UMTS/HSPA+ femtocells," *IEEE Commun. Mag.*, vol. 47, no. 9, pp. 102–109, Sep. 2009.
3. J.-H. Yun and K. Shin, "Adaptive interference management of OFDMA femtocells for co-channel deployment," *IEEE J. Select. Areas Commun.*, vol. 29, no. 6, pp. 1225–1241, Jun. 2011.
4. D. T. Ngo, L. B. Le, T. Le-Ngoc, E. Hossain, and D. I. Kim, "Distributed interference management in femtocell networks," in *Proc. IEEE Veh. Tech. Conf. (VTC-Fall)*, San Franciso, CA, Sep. 2011, pp. 1–5.
5. D. T. Ngo, L. B. Le, T. Le-Ngoc, E. Hossain, and D. I. Kim, "Distributed interference management in two-tier CDMA femtocell networks," *IEEE Trans. Wireless Commun.*, vol. 11, no. 3, pp. 979–989, Mar. 2012.
6. D. T. Ngo, L. B. Le, and T. Le-Ngoc, "Distributed Pareto-optimal power control in femtocell networks," in *Proc. IEEE Intl. Symp. on Personal, Indoor and Mobile Radio Commun. (PIMRC)*, Toronto, ON, Canada, Sep. 2011, pp. 222–226.
7. D. T. Ngo, L. B. Le, and T. Le-Ngoc, "Joint utility maximization in two-tier networks by distributed Pareto-optimal power control," in *Proc. IEEE Vehicular Technology Conf. (VTC-Fall)*, Quebec City, QC, Canada, Sep. 2012, pp. 1–5.
8. D. T. Ngo, L. B. Le, and T. Le-Ngoc, "Distributed Pareto-optimal power control for utility maximization in femtocell networks," *IEEE Trans. Wireless Commun.*, vol. 11, no. 10, pp. 3434–3446, Oct. 2012.
9. D. Lopez-Perez, A. Valcarce, G. de la Roche, and J. Zhang, "OFDMA femtocells: A roadmap on interference avoidance," *IEEE Commun. Mag.*, vol. 47, no. 9, pp. 41–48, Sep. 2009.
10. N. Saquib, E. Hossain, L. B. Le, and D. I. Kim, "Interference management in OFDMA femtocell networks: Issues and approaches," *IEEE Wirel. Commun.*, vol. 19, no. 3, pp. 86–95, Jun. 2012.
11. D. T. Ngo, S. Khakurel, and T. Le-Ngoc, "Distributed subchannel and power allocation for OFDMA-based femtocell networks," in *IEEE Vehicular Technology Conf. (VTC-Spring)*, Dresden, Germany, Jun. 2013, pp. 1–5.
12. D. T. Ngo, S. Khakurel, and T. Le-Ngoc, "Joint subchannel assignment and power allocation for OFDMA femtocell networks," *IEEE Trans. Wireless Commun.*, vol. 13, no. 1, pp. 342–355, Jan. 2014.
13. B. R. Marks and G. P. Wright, "A general inner approximation algorithm for nonconvex mathematical programs," *Operations Research*, vol. 26, no. 4, pp. 681–683, 1978.
14. S. Boyd and L. Vandenberghe, *Convex optimization*. Cambridge University Press, 2004.
15. J. Papandriopoulos and J. S. Evans, "SCALE: A low-complexity distributed protocol for spectrum balancing in multiuser DSL networks," *IEEE Trans. Inform. Theory*, vol. 55, no. 8, pp. 3711–3724, Aug. 2009.
16. H. D. Tuan, S. Hosoe, and H. Tuy, "D.C. optimization approach to robust controls: The optimal scaling value problem," *IEEE Trans. Automatic Control*, vol. 45, pp. 1903–1909, Sep. 2000.
17. M. Frank and P. Wolfe, "An algorithm for quadratic programming," *Naval Research Logist. Quart.*, vol. 3, pp. 95–110, 1956.
18. M. Chiang, C. W. Tan, D. P. Palomar, D. O'Neill, and D. Julian, "Power control by geometric programming," *IEEE Trans. Wireless Commun.*, vol. 6, no. 7, pp. 2640–2651, Jul. 2007.
19. H. H. Kha, H. D. Tuan, and H. H. Nguyen, "Fast global optimal power allocation in wireless networks by local D.C. programming," *IEEE Trans. Wireless Commun.*, vol. 11, no. 2, pp. 510–515, Feb. 2012.
20. L. Venturino, N. Prasad, and X. Wang, "Coordinated scheduling and power allocation in downlink multicell OFDMA networks," *IEEE Trans. Veh. Technol.*, vol. 58, no. 6, pp. 2835–2848, Jul. 2009.

21. Y. Liu and E. Knightly, "Opportunistic fair scheduling over multiple wireless channels," in *IEEE INFOCOM*, San Francisco, CA, Mar. 2003, pp. 1106–1115.
22. T. Wang and L. Vandendorpe, "Iterative resource allocation for maximizing weighted sum minrate in downlink cellular OFDMA systems," *IEEE Trans. Signal Processing*, vol. 59, no. 1, pp. 223–234, Jan. 2011.
23. K. Son, S. Lee, Y. Yi, and S. Chong, "REFIM: A practical interference management in heterogeneous wireless access networks," *IEEE J. Select. Areas Commun.*, vol. 29, no. 6, pp. 1260–1272, Jun. 2011.
24. D. P. Bertsekas, *Nonlinear Programming*, 2nd ed. Boston: Athena Scientific, 1999.
25. P. Hande, S. Rangan, M. Chiang, and X. Wu, "Distributed uplink power control for optimal SIR assignment in cellular data networks," *IEEE/ACM Trans. Netw.*, vol. 16, no. 6, pp. 1420–1433, Dec. 2008.
26. V. Chandrasekhar, J. G. Andrews, T. Muharemovic, and Z. Shen, "Power control in two-tier femtocell networks," *IEEE Trans. Wireless Commun.*, vol. 8, no. 8, pp. 4316–4328, Aug. 2009.
27. R. Cendrillon, J. Huang, M. Chiang, and M. Moonen, "Autonomous spectrum balancing for digital subscriber lines," *IEEE Trans. Signal Processing*, vol. 55, no. 8, pp. 4241–4257, Aug. 2007.

Chapter 6
Distributed Resource Allocation in OFDMA Cognitive Small-Cell Networks with Spectrum-Sharing Constraints

Several recent studies have proposed that heterogeneous small-cell networks can adopt the model of cognitive radio, in that newly deployed FUEs opportunistically access the radio spectrum already licensed to MUEs [1–3]. In this model, MUEs play the role of PUs whereas FUEs play the role of SUs. Different from Chap. 5 where cochannel deployment is assumed for both macrocell and femtocells [4, 5], this chapter proposes distributed algorithms for OFDMA-based cognitive femtocell networks with opportunistic spectrum access [6, 7]. As in any cognitive radio network, the key design challenge here is to protect MUEs from excessive interference induced by FUEs whilst providing the latter with some QoS [8, 9].

In cognitive heterogeneous networks where MUEs are highly dynamic on the radio spectrum and the opportunity for secondary access is small, it is critical to efficiently share the temporarily available frequency bands among the FUEs. The fair allocation of radio resources in OFDMA-based systems has been previously studied in different contexts, including max-min fairness [10], proportional rate guarantee [11], minimum bandwidth assurance [12, 13], equal bandwidth distribution [14], and proportional fairness [15]. However, none of these solutions accounts for a flexible way to control the sharing of radio spectrum in such a dynamic access scenario.

Apart from the tolerable interference limits at MUEs, the problems formulated in this chapter also include upper and lower bounds on the number of subchannels that FUEs are allowed to use. While these limits specify the priority in terms of spectrum access opportunity provided to FUEs, a notion of spectrum-sharing fairness is also indicated. For instance, the upper limits can be used to prevent the FUEs with favorable channel conditions from greedily filling up all the spectrum holes. On the other hand, the lower thresholds guarantee that a minimum bandwidth is given to other FUEs. Fairness in bandwidth allocation can be translated into that of attainable throughput, because there are direct relations between these two parameters.

These additional constraints imply a new dimension of technical difficulty in finding efficient solutions for the optimization problems at hand. The popular winner-take-all approach [16, 17], which always allocates radio resources

D.T. Ngo and T. Le-Ngoc, *Architectures of Small-Cell Networks and Interference Management*, SpringerBriefs in Computer Science, DOI 10.1007/978-3-319-04822-2_6, © The Author(s) 2014

to the most advantageous FUEs, is no longer applicable. This work applies Lagrangian dual optimization to devise globally optimal solutions for both sum-rate maximization and power minimization problems. It is shown that the complexity of the proposed solutions only grows polynomially in the number of OFDM subchannels, compared to an exponential complexity typically required by direct search methods. From the dual framework, the devised solutions are implemented by distributed algorithms, which do not require any central coordination. It is proposed that the limited network cooperation is accomplished by exchanging relevant information over a common reserved channel and by implementing virtual timers at each FUE.

6.1 System Model and Problem Formulation

We consider a scenario in which a macrocell (primary) BS transmits M *downlink* traffic flows (not necessarily OFDM) to its M subscribed MUEs. Denote the set of all MUEs by $\mathscr{L}_m = \{1, \ldots, M\}$. Assume that each of those data streams is intended for only one MUE and occupies a predetermined frequency band $B_m, m \in \mathscr{L}_m$ in the radio spectrum. As the macrocell network does not utilize the entire available spectral ranges, a femtocell network is deployed to implement efficient opportunistic spectrum access. This secondary network, which consists of K transmitter (Tx)–receiver (Rx) pairs, is assumed to be capable of accurately sensing the spectrum to locate the frequency bands temporarily unused by the macrocell network. Denote the set of all FUE pairs by $\mathscr{L}_f = \{1, \ldots, K\}$. Each of these femtocell Tx-Rx pairs can model an FUE-femtocell BS connection, e.g., in the uplink the Tx is the FUE whereas in the downlink it is the femtocell BS. The set of all UEs is $\mathscr{L} = \mathscr{L}_m \cup \mathscr{L}_f$.

The spectral holes are merged into a common pool according to the spectrum pooling approach, from which the total bandwidth B available for secondary access is divided into N OFDM subchannels of equal bandwidth $B_N = B/N$. Denote by $\mathscr{N} = \{1, \ldots, N\}$ the set of available OFDM subchannels. Also let $\mathscr{N}_k$ represent the set of OFDM subchannels allocated to FUE $k \in \mathscr{L}_f$, and $|\mathscr{N}_k|$ is its cardinality. An example of the considered system is depicted in Fig. 6.1. Although macrocell and femtocell networks are supposed not to occupy the same frequency bands at the same time, the non-orthogonality of their transmitted signals may lead to mutual interference, as shown in the followings.

Denote the channel gain from the transmitting side of UE $l \in \mathscr{L}$ to the receiving side of UE $l' \in \mathscr{L}$ on subchannel n as $h_{l,l'}^{(n)}$. Let $\Phi_m(e^{jw})$ be the power spectral density (PSD) of the signal transmitted by the macrocell BS to MUE $m \in \mathscr{L}_m$ on frequency B_m. The interference caused by this signal to subchannel $n \in \mathscr{N}$ is [18]:

$$J_m^{(n)} = \frac{1}{2\pi N} \int_{\bar{d}_m^{(n)} - B_N/2}^{\bar{d}_m^{(n)} + B_N/2} \int_{-\pi}^{\pi} \Phi_m(e^{jw}) \left(\frac{\sin\left[\frac{N}{2}(w - \phi)\right]}{\sin\left[\frac{1}{2}(w - \phi)\right]} \right)^2 d\phi \, dw, \qquad (6.1)$$

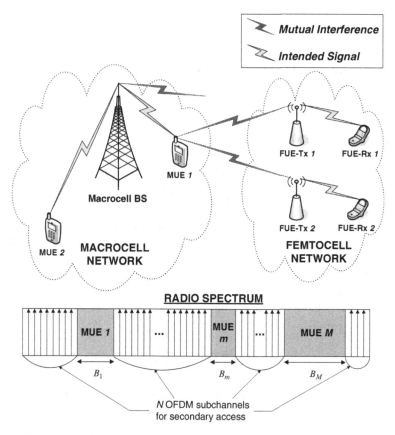

Fig. 6.1 Coexistence of a macrocell (primary) network and a cognitive femtocell (secondary) network

where $\bar{d}_m^{(n)} = |f_n - f_m|$ represents the spectral distance between subchannel $n \in \mathcal{N}$ and the center frequency f_m of MUE $m \in \mathcal{L}_m$. The total interference introduced by the macrocell BS to femto-Rx $k \in \mathcal{L}_f$ on $n \in \mathcal{N}$ can be expressed as:

$$\check{J}_k^{(n)} = h_{m,k}^{(n)} \sum_{m \in \mathcal{L}_m} J_m^{(n)}. \tag{6.2}$$

On the other hand, the OFDM signals from a femto-Tx to its intended femto-Rx might interfere the reception at the MUEs. Denote T_s as the OFDM symbol duration and define $p_k^{(n)}$ as the transmit power from femto-Tx $k \in \mathcal{L}_f$ to femto-Rx $k \in \mathcal{L}_f$ on subchannel $n \in \mathcal{N}$. Let $\mathbf{p}_k = [p_k^{(1)}, p_k^{(2)}, \ldots, p_k^{(N)}]^T$ and $\mathbf{p} = \mathrm{vec}[\mathbf{p}_1, \mathbf{p}_2, \ldots, \mathbf{p}_K]$. The PSD of the subchannel-n signal by FUE $k \in \mathcal{L}_f$ can be modeled as:

$$\Phi_k^{(n)}(f) = p_k^{(n)} T_s \left(\frac{\sin \pi f T_s}{\pi f T_s} \right)^2. \tag{6.3}$$

The interference caused by this signal onto MUE $m \in \mathscr{L}_m$ is then [18]:

$$I_{k,m}^{(n)} = p_k^{(n)} \check{I}_{k,m}^{(n)}, \tag{6.4}$$

where $\check{I}_{k,m}^{(n)}$ is defined as:

$$\check{I}_{k,m}^{(n)} = h_{k,m}^{(n)} T_s \int_{\bar{d}_m^{(n)} - B_m/2}^{\bar{d}_m^{(n)} + B_m/2} \left(\frac{\sin \pi f T_s}{\pi f T_s} \right)^2 df. \tag{6.5}$$

In this work, we assume a slow fading channel model such that the channel conditions remain unchanged during the resource allocation period (for instance, in high data rate systems and/or environments with reduced degrees of mobility) [10, 16, 19]. This is the case when the coherence time of the channel is larger than the symbol period of the transmitted signal [20]. The channel-to-interference-plus-noise ratio (CINR) of FUE $k \in \mathscr{L}_f$ on subchannel $n \in \mathscr{N}$ can be written as:

$$\gamma_k^{(n)} = \frac{h_{k,k}^{(n)}}{\check{J}_k^{(n)} + B_N N_0}, \tag{6.6}$$

where N_0 is the one-sided PSD of AWGN. The maximum attainable rate of FUE $k \in \mathscr{L}_f$ on subchannel $n \in \mathscr{N}$ can be expressed as:

$$r_k^{(n)} = \ln \left(1 + \gamma_k^{(n)} p_k^{(n)} \right), \tag{6.7}$$

and its the sum rate is

$$r_k = \sum_{n \in \mathscr{N}} r_k^{(n)} = \sum_{n \in \mathscr{N}} \ln \left(1 + \gamma_k^{(n)} p_k^{(n)} \right). \tag{6.8}$$

One of the main goals of this work is to devise a joint subchannel assignment and power allocation scheme that maximizes the aggregate throughput of all secondary transmissions. At the same time, we target to satisfy important constraints on the maximum tolerable interference at each MUE and on the total transmit powers of individual femto-Tx's. It should be emphasized that since MUEs always have priority access to the radio spectrum, the chances for cognitive FUEs to utilize the frequencies depend heavily on the dynamics of these licensed users. In cases where MUEs are extremely active and occupy many frequency bands for a long period of time, the actual opportunities left for secondary access become rare. Therefore, it is critical to share out these valuable but yet scarce frequency bands among the FUEs in an efficient manner. To this end, we propose to constrain the maximum and minimum numbers of OFDM subchannels that individual femtocell Tx-Rx pairs are permitted to use. Specifically, the design problem is formulated as follows:

$$\max_{\mathbf{p}} \sum_{k\in\mathscr{L}_f} w_k \sum_{n\in\mathscr{N}_k} \ln\left(1 + \gamma_k^{(n)} p_k^{(n)}\right) \tag{6.9a}$$

$$\text{s.t.} \sum_{k\in\mathscr{L}_f} \sum_{n\in\mathscr{N}_k} p_k^{(n)} \check{I}_{k,m}^{(n)} \le I_m^{\text{th}}, \quad \forall m \in \mathscr{L}_m \tag{6.9b}$$

$$\sum_{n\in\mathscr{N}_k} p_k^{(n)} \le P_k^{\max}, \quad \forall k \in \mathscr{L}_f \tag{6.9c}$$

$$p_k^{(n)} \ge 0, \quad \forall k \in \mathscr{L}_f, \forall n \in \mathscr{N}_k \tag{6.9d}$$

$$p_k^{(n)} p_{k'}^{(n)} = 0, \quad \forall n \in \mathscr{N}, \forall k' \neq k \in \mathscr{L}_f \tag{6.9e}$$

$$N_k^{\min} \le |\mathscr{N}_k| \le N_k^{\max}, \quad \forall k \in \mathscr{L}_f. \tag{6.9f}$$

In the above formulation, the fixed weight vector $\mathbf{w} = [w_1, w_2, \ldots, w_K]^T$ is required to satisfy $0 \le w_k \le 1$ and $\sum_{k\in\mathscr{L}_f} w_k = 1$. A larger value of w_k represents a higher priority given to FUE $k \in \mathscr{L}_k$. With I_m^{th} denoting the interference threshold, constraint (6.9b) expresses the maximum allowable interference at MUE $m \in \mathscr{L}_m$. The regulatory limit on the total transmit power at femto-Tx $k \in \mathscr{L}_k$ is represented in (6.9c), whereas (6.9d) and (6.9e) enforce a disjoint subchannel assignment in OFDMA systems, i.e., one subchannel is assigned to at most one FUE at a time.

In particular, the spectrum-sharing[1] constraints are expressed in (6.9f) where the total number of subchannels allotted to any FUE $k \in \mathscr{L}_k$ is upper- and lower-bounded by $N_k^{\max}$ and $N_k^{\min}$, respectively. These two limits can be used to specify the priority in terms of spectrum access opportunity granted to individual FUEs. As $N_k^{\max}$ and $N_k^{\min}$ increase toward N, a higher priority is given to femtocell Tx-Rx pair $k \in \mathscr{L}_k$. Certain notion of bandwidth-sharing fairness can also be realized with (6.9f). For instance, by giving lower values of $N_k^{\max}$'s to the FUEs with favorable conditions, it is possible to effectively prevent such users from greedily taking up most of the valuable secondary spectrum. Other FUEs can be assigned higher values of $N_k^{\min}$'s so as to meet some QoS requirements. If however $N_k^{\max} = N_k^{\min} = \lfloor N/K \rfloor$ for all $k \in \mathscr{L}_f$, the optimization formulation enforces (almost) a strict bandwidth fairness. In many cases, the fairness in terms of bandwidth can be translated into that of attainable throughput. Specific arrangement of $N_k^{\max}$ and $N_k^{\min}$ values, in practice, can be reached a priori (i) upon agreement among FUEs or (ii) by a secondary spectrum authority. Nevertheless, these two limits are strictly required to satisfy $\sum_{k\in\mathscr{L}_f} N_k^{\min} \le N$ and $N_k^{\max} \le N$, $\forall k \in \mathscr{L}_f$.

---

[1]Within the scope of this chapter, "spectrum-sharing" means sharing of the temporarily available radio frequencies among FUEs.

It is noteworthy that solving the optimization problem (6.9) is challenging since it requires the allocation of an optimal set of subchannels to each femto Tx-Rx pair. As noted in [21], the computational complexity needed to directly resolve this type of combinatorial problems increases, at least, exponentially with the number of subchannels N. The new constraints (6.9f) even pose more challenges in finding efficient solutions to (6.9). Indeed, the popular winner-take-all policies available in the literature (see, e.g., [16, 17]) can be invalid to the problem at hand. Meanwhile, cognitive radio applications demand optimal solutions to be delivered in a timely fashion to cope with the quick changes of wireless environments.

6.2 Joint Subchannel and Power Allocation for Throughput Maximization in Cognitive Femtocell Networks

Motivated by the aforementioned observations, we will develop an optimal algorithm to efficiently resolve problem (6.9) where, instead, the solution is derived in the dual domain. The main motivation behind this approach is that the particular structure of (6.9) satisfies the so-called "frequency-sharing" condition [22, Theorem 1]. In this case, the dual-domain optimal subchannel-power allocation will become that of the primal problem for a sufficiently large number of subchannels.

6.2.1 Optimal Design with Spectrum-Sharing Constraints

Excluding (6.9e) and (6.9f), the Lagrangian of (6.9) is defined as:

$$\mathcal{L}(\mathbf{p}, \boldsymbol{\lambda}, \boldsymbol{\mu}) = \sum_{k \in \mathcal{L}_f} w_k \sum_{n \in \mathcal{N}_k} \ln\left(1 + \gamma_k^{(n)} p_k^{(n)}\right) - \sum_{m \in \mathcal{L}_m} \lambda_m \left(\sum_{k \in \mathcal{L}_f} \sum_{n \in \mathcal{N}_k} p_k^{(n)} \check{I}_{k,m}^{(n)} - I_m^{\text{th}}\right)$$

$$- \sum_{k \in \mathcal{L}_f} \mu_k \left(\sum_{n \in \mathcal{N}_k} p_k^{(n)} - P_k^{\max}\right), \tag{6.10}$$

where $\boldsymbol{\lambda} = [\lambda_1, \ldots, \lambda_M]^T \succeq \mathbf{0}$ and $\boldsymbol{\mu} = [\mu_1, \ldots, \mu_K]^T \succeq \mathbf{0}$ are the dual variables. The Lagrange dual function is thus:

$$D(\boldsymbol{\lambda}, \boldsymbol{\mu}) = \max_{\mathbf{p} \succeq \mathbf{0}} \mathcal{L}(\mathbf{p}, \boldsymbol{\lambda}, \boldsymbol{\mu}). \tag{6.11}$$

By appropriately swapping the order in the summations and using (6.9e), (6.11) is decomposed into N independent problems, one for each subchannel $n \in \mathcal{N}$ as:

$$D(\boldsymbol{\lambda}, \boldsymbol{\mu}) = \sum_{n \in \mathcal{N}} D^{(n)}(\boldsymbol{\lambda}, \boldsymbol{\mu}) + \sum_{m \in \mathcal{L}_m} \lambda_m I_m^{\text{th}} + \sum_{k \in \mathcal{L}_f} \mu_k P_k^{\max}, \tag{6.12}$$

with the per-subchannel optimization problem being

$$D^{(n)}(\boldsymbol{\lambda}, \boldsymbol{\mu}) = \max_{\mathbf{p} \succeq 0} \sum_{k \in \mathcal{L}_f} \left\{ w_k \ln \left(1 + \gamma_k^{(n)} p_k^{(n)} \right) - \left(\sum_{m \in \mathcal{L}_m} \lambda_m \check{I}_{k,m}^{(n)} + \mu_k \right) p_k^{(n)} \right\}.$$
$$\tag{6.13}$$

For each subchannel $n \in \mathcal{N}$, there is at most one $p_k^{(n)} > 0, \forall k \in \mathcal{L}_f$. Assume that FUE k is active on subchannel n. Given a fixed $(\boldsymbol{\lambda}, \boldsymbol{\mu})$, it can be shown that the objective of the maximization in (6.13) is a concave function with respect to $p_k^{(n)}$. From the KKT conditions [23], the optimal power allocation can be devised as:

$$p_k^{(n)*} = \left(\frac{w_k}{\sum_{m \in \mathcal{L}_m} \lambda_m \check{I}_{k,m}^{(n)} + \mu_k} - \frac{1}{\gamma_k^{(n)}} \right)^+. \tag{6.14}$$

Substituting (6.14) to (6.13), one obtains K possible values of $D^{(n)}(\boldsymbol{\lambda}, \boldsymbol{\mu})$ as:

$$D_k^{(n)}(\boldsymbol{\lambda}, \boldsymbol{\mu}) = w_k \ln \left(1 + \gamma_k^{(n)} p_k^{(n)*} \right) - \left(\sum_{m \in \mathcal{L}_m} \lambda_m \check{I}_{k,m}^{(n)} + \mu_k \right) p_k^{(n)*}, \quad \forall k \in \mathcal{L}_f. $$
$$\tag{6.15}$$

Notice that $\mathbf{p}$ is also required to satisfy the spectrum-sharing constraints (6.9f). For a particular subchannel $n \in \mathcal{N}$, the task of optimally determining which FUE to use n *cannot* be done by simply selecting among the total K power allocations from (6.15) the one that maximizes $D^{(n)}(\boldsymbol{\lambda}, \boldsymbol{\mu})$. Instead, this involves searching through all KN values of $D_k^{(n)}(\boldsymbol{\lambda}, \boldsymbol{\mu})$ to decide the optimal subchannel-FUE matchings and subsequently the optimal power distributions for those assignments.

Towards this end, we propose a two-phase procedure that designates $p_k^{(n)*}, k \in \mathcal{L}_f, n \in \mathcal{N}$ to their eligible FUEs in an optimal fashion while also satisfying (6.9f). The main steps of this procedure are outlined in Table 6.1. Specifically, Phase 1 attempts to provide minimum guarantee on the number of subchannels allotted to all FUEs. Then, to further enhance the system throughput the remaining subchannels are allocated on a competitive basis among the FUEs in Phase 2. Effectively, the proposed procedure ensures that all the subchannels are fully occupied by the cognitive femtocell network. In any case, as soon as a certain FUE has reached its maximum allowable shares of spectrum, it will be eliminated from the competition until the end of the allocation round (for these fixed values of λ and μ).

Table 6.1 Joint allocation of subchannels and power for a fixed $\{\boldsymbol{\lambda}, \boldsymbol{\mu}\}$

Initialization

- Given $\boldsymbol{\lambda}, \boldsymbol{\mu}$, compute all KN power allocation by (6.14).
- Construct $\mathbf{A} := \left[D_k^{(n)}(\boldsymbol{\lambda}, \boldsymbol{\mu}) \right]_{k,n} \in \mathbb{R}^{K \times N}$ as in (6.15).

PHASE 1 – Allocation to meet spectrum-sharing constraints

- If $N_k^{\min} = 0$, discard FUE $k \in \mathscr{L}_f$ from further consideration in Phase 1.

Repeat

   ◦ Perform a 2-D search on $\mathbf{A}$ to find the entry $\mathbf{A}(\bar{k}, \bar{n})$ with the largest value.

   ◦ Assign $p_{\bar{k}}^{\bar{n}} := p_{\bar{k}}^{\bar{n}*}$ and $p_k^{\bar{n}} := 0, \forall k \neq \bar{k} \in \mathscr{L}_f$.

   ◦ Discard subchannel $\bar{n}$ from all subsequent searches.

   ◦ If $|\mathscr{N}_{\bar{k}}| = N_{\bar{k}}^{\max}$, discard FUE $\bar{k}$ from all subsequent searches.

   ◦ If $|\mathscr{N}_{\bar{k}}| = N_{\bar{k}}^{\min}$, discard FUE $\bar{k}$ in the rest of Phase 1.

Until $|\mathscr{N}_k| \geq N_k^{\min}, \; \forall k \in \mathscr{L}_f$

PHASE 2 – Allocation to further enhance system throughput

Repeat

   ◦ Perform a 2-D search on $\mathbf{A}$ to find the entry $\mathbf{A}(\bar{k}, \bar{n})$ with the largest value.

   ◦ Assign $p_{\bar{k}}^{\bar{n}} := p_{\bar{k}}^{\bar{n}*}$ and $p_k^{\bar{n}} := 0, \forall k \neq \bar{k} \in \mathscr{L}_f$.

   ◦ Discard subchannel $\bar{n}$ from all subsequent searches.

   ◦ If $|\mathscr{N}_{\bar{k}}| = N_{\bar{k}}^{\max}$, discard FUE $\bar{k}$ from all subsequent searches.

Until All subchannels $n \in \mathscr{N}$ are assigned.

Once the subchannel-power allocation has been established, solutions to all N unconstrained optimization problems (6.13) are readily available. As such, the overall Lagrange dual function $D(\boldsymbol{\lambda}, \boldsymbol{\mu})$ in (6.12) can be evaluated for the fixed $(\boldsymbol{\lambda}, \boldsymbol{\mu})$. It now remains to solve:

$$\min_{\boldsymbol{\lambda} \geq 0, \, \boldsymbol{\mu} \geq 0} D(\boldsymbol{\lambda}, \boldsymbol{\mu}). \tag{6.16}$$

This Lagrange dual problem is convex, regardless of the convexity of the primal problem (6.9). Its solution can thus be determined by a subgradient method, which iteratively updates $\{\boldsymbol{\lambda}, \boldsymbol{\mu}\}$ in the subgradient direction until convergence. For our problem of interest, the updates may be performed as:

$$\lambda_m[t+1] = \left(\lambda_m[t] - \delta_\lambda \left[I_m^{\text{th}} - \sum_{k \in \mathscr{L}_f} \sum_{n \in \mathscr{N}_k} p_k^{(n)*} \breve{I}_{k,m}^{(n)} \right] \right)^+, \quad m \in \mathscr{L}_m \tag{6.17}$$

$$\mu_k[t+1] = \left(\mu_k[t] - \delta_\mu \left[P_k^{\max} - \sum_{n \in \mathscr{N}_k} p_k^{(n)*} \right] \right)^+, \quad k \in \mathscr{L}_f \tag{6.18}$$

with sequences of scalar step sizes $\delta_\lambda, \delta_\mu > 0$. As long as δ_λ and δ_μ are chosen to be sufficiently small, these subgradient updates are guaranteed to converge, resulting in optimal $(\boldsymbol{\lambda}^*, \boldsymbol{\mu}^*)$ [23]. Some popular choices include the constant step size rule

c, and diminishing rules c/t and $c/\sqrt{t}$ for some constant $c > 0$. At the point of convergence, together with the dual optimal solutions $(\boldsymbol{\lambda}^*, \boldsymbol{\mu}^*)$, the primal-domain optimal allocation $p_k^{(n)*}$ can also be recovered.

As noted in [22, Theorem 1] and [21, Sect. III-C], the optimal value of the original optimization problem (6.9) is exactly the minimal value of $D(\boldsymbol{\lambda}, \boldsymbol{\mu})$ in (6.16), provided that N is sufficiently large. Depending on specific system parameters (e.g., channel gains and mutual interferences), there are cases in which the solution that gives this globally optimal value is not unique. An example is when there are equal values of $D_k^{(n)}$'s in (6.15), implying multiple possible optimal subchannel-FUE matchings. This corresponds to the existence of many equivalent optimal allocation solutions.

It is worth pointing out that the above derived procedure is of centralized nature, since at least a 2-D search is necessary to determine the subchannel and power allocation for all $k \in \mathscr{L}_f$ and $n \in \mathscr{N}$ [refer to Table 6.1]. As a consequence, there should be a central processing unit that collects all the information regarding the network states, performs the optimization procedures, and disseminates its computation results to the corresponding network entities. On one hand, these communication overheads may cause intolerable delays in environments with dynamic MUE activities, where any optimizing solutions need to be found quickly while secondary access is still possible. On the other hand, it can be challenging to implement such centralized coordination in a femtocell networking scenario. Based upon the structure of solutions obtained through Lagrangian duality, we now propose a distributed design that only requires some limited cooperation among macrocell and femtocell networks while offering optimal performance.

6.2.2 Distributed Implementation

The keys to realize a distributed solution for (6.9) lie in the search for optimal $p_k^{(n)*}$ to solve (6.13) while also satisfying constraint (6.9f). Observe that the computation of $p_k^{(n)*}$ in (6.14) mainly requires the *local* information available at femto-Tx's k itself, except for λ_m and $\check{I}_{k,m}^{(n)}$. Since both $h_{k,k}^{(n)}$ and $\check{J}_k^{(n)}$ can be estimated/measured at femto-Rx k, CINR $\gamma_k^{(n)}$ can be computed and made available to its corresponding transmitter via a dedicate channel (to be discussed later). To evaluate $\lambda_m \check{I}_{k,m}^{(n)}$, femto-Tx $k \in \mathscr{L}_f$ demands certain collaboration from MUE $m \in \mathscr{L}_m$. Specifically, each MUE m broadcasts the updated λ_m and estimated channel gain $h_{k,m}^{(n)}$ to FUE k. Upon receiving these values from all M MUEs, femto-Tx k is able to determine $p_k^{(n)*}$ and $D_k^{(n)}(\boldsymbol{\lambda}, \boldsymbol{\mu}), \forall n \in \mathscr{N}$ by (6.15).

The above-proposed procedure in Table 6.1 can be implemented in a distributive fashion as follows. At the beginning of the allocation period, individual femto-Tx's broadcast "MIN-SUB-REQ" flag packets to specify their required minimum number of subchannels $N_k^{\min}$. Upon receiving all these packets and by summing the indicated values together, femto-Tx's know that $N_{\text{tot}}^{\min}$ subchannels are requested. Then, based on the computed $p_k^{(n)*}$ and $D_k^{(n)}(\lambda, \mu)$, each femto-Tx $k \in \mathscr{L}_f$ constructs a length-N list of virtual timers $\mathbf{T}_k$:

$$T_k^{(n)} = c_T \exp\left[-D_k^{(n)}(\lambda, \mu)\right], \quad \forall n \in \mathscr{N}, \tag{6.19}$$

where constant $c_T > 0$ is made common to all FUEs. Notice that our definition of virtual timer in (6.19) is able to deal with any real-valued $D_k^{(n)}(\lambda, \mu)$. This is rather different from the "virtual clock" values defined in [16], only valid for strictly positive $D_k^{(n)}(\lambda, \mu)$.

As there are N OFDM subchannels available for secondary access, we divide the total allocation time into N minislots and let FUEs compete in a sequential manner. At the beginning of competing minislot i, all femto-Tx $k \in \mathscr{L}_f$ whose $N_k^{\min} > 0$ pick the *largest* value from their own list $\mathbf{T}_k$ and start the virtual timer corresponding to that value. Assume that the largest value $D_{\bar{k}}^{(n^{(i)})}(\lambda, \mu)$ of femto-Tx $\bar{k}$ is also the maximum over all FUEs' virtual timer lists. The timer of $\bar{k}$ expires first[2], and it is allowed to transmit with power $p_{\bar{k}}^{(n^{(i)})*}$ over subchannel $n^{(i)}$. FUE $\bar{k}$ then broadcasts an "EXPIRE" flag packet to indicate that subchannel $n^{(i)}$ has already been occupied. Since one subchannel can be used by at most one FUE, upon receiving this message all other femto-Tx's remove the entry corresponding to subchannel $n^{(i)}$ from their lists of virtual timers. Based on the counters of these "EXPIRE" messages and the computed value of $N_{\text{tot}}^{\min}$, femto-Tx's know if all other FUEs have been assigned with their minimum number of subchannels, and switch between "WAITING" and "COMPETING" modes accordingly. By counting the "EXPIRE" packets, both femto-Tx's and MUEs are also aware of when to update their respective Lagrangian variables μ_k's and λ_m's.

The above implementation, which shall be referred to as Distributed Throughput-Maximization with Spectrum-sharing Constraints (D-TMSC) scheme, is described in Tables 6.2 and 6.3. This decentralized protocol features the autonomous operation of individual users in both cognitive femtocell and macrocell networks, with a limited level of cooperation required. Here, we assume that the exchange of messages is completed without errors via a dedicated control channel.

- *Within the cognitive femtocell network*:

  1. Femto-Rx k feedbacks its estimate of CINR $\gamma_k^{(n)}$ to femto-Tx k.
  2. Femto-Tx's have two operating modes: "WAITING" and "COMPETING."
  3. Two types of flag messages are broadcast by femto-Tx's: "MIN-SUB-REQ" and "EXPIRE" packets.

---

[2]If two or more timers simultaneously expire, all of the collided FUEs back off and generate their own random timers (e.g., using the uniform distribution). The one that expires first is the winner of that minislot. This approach is similar to CSMA [24].

Table 6.2 Proposed D-TMSC algorithm – At the cognitive femtocell network

---

PHASE 1 – Initialization

• Each femto-Tx $k \in \mathscr{L}_f$ broadcasts a "MIN-SUB-REQ" flag packet to indicate its required minimum number of subchannels $N_k^{\min}$. Upon receiving all of these "MIN-SUB-REQ" packets, femto-Tx's know that $N_{\text{tot}}^{\min}$ subchannels are requested. Each femto-Tx k computes all $p_k^{(n)*}$'s [by (6.14)] and constructs a list of virtual timers $\mathbf{T}^{(k)}$ [by (6.15) and (6.19)]. Femto-Tx k initializes μ_k, and sets $t := 0, i := 1$.

PHASE 2 – Subchannel-power allocation for minimum-bandwidth guarantee

• Femto-Tx k with $N_k^{\min} = 0$ switches to "WAITING" mode and remains in that mode until receiving $N_{\text{tot}}^{\min}$ "EXPIRE" flag packets. Other femto-Tx's switch to "COMPETING" mode and start their respective largest virtual timers for this minislot i.

• Femto-Tx expiring first (denoted as $\bar{k}$) is eligible to use subchannel $n^{(i)}$. It then broadcasts an "EXPIRE" flag packet to indicate that subchannel $n^{(i)}$ has been occupied. Upon receiving this packet, other femto-Tx's stop their virtual timers, back off, delete the entry corresponding to $n^{(i)}$ from their virtual timer lists, and move to time slot $i := i + 1$.

• Femto-Tx $\bar{k}$ starts transmitting with power $p_{\bar{k}}^{(n^{(i)})*}$ on subchannel $n^{(i)}$.

• If femto-Tx $\bar{k}$ recognizes that $|N_{\bar{k}}| = N_{\bar{k}}^{\max}$, it will switch to and remain in "WAITING" mode until receiving N "EXPIRE" packets (including those generated by itself). If femto-Tx $\bar{k}$ realizes that $|N_{\bar{k}}| = N_{\bar{k}}^{\min}$, it will switch to and remain in "WAITING" mode until receiving $N_{\min}^{\text{tot}}$ "EXPIRE" flag packets (including those generated by itself). Otherwise, femto-Tx $\bar{k}$ deletes the entry corresponding to subchannel $n^{(i)}$ from its virtual timer list, and moves to time slot $i := i + 1$.

PHASE 3 – Subchannel-power allocation to enhance throughput

• Femto-Tx's switch to "COMPETING" mode, start their largest virtual timers for minislot i.

• Femto-Tx expiring first (denoted as $\bar{k}$) is eligible to occupy this subchannel $n^{(i)}$. It then broadcasts an "EXPIRE" packet. Upon receiving this packet, other femto-Tx's stop their virtual timers, back off, delete the entry corresponding to subchannel $n^{(i)}$ from their virtual timer lists, and move to time slot $i := i + 1$.

• Femto-Tx $\bar{k}$ starts transmitting with power $p_{\bar{k}}^{(n^{(i)})*}$ on subchannel $n^{(i)}$.

• If femto-Tx $\bar{k}$ recognizes that $|N_{\bar{k}}| = N_{\bar{k}}^{\max}$, it will switch to and remain in "WAITING" mode until receiving N "EXPIRE" packets (including those generated by itself).

PHASE 4 – Lagrangian update

• Upon receiving all N "EXPIRE" packets, each femto-Tx $k \in \mathscr{L}_f$ updates its own μ_k based on (6.18), resets its "EXPIRE" message counter, sets $t := t + 1$, and resets $i := 1$.

• Return to Phase 2 and repeat until convergence.

Table 6.3 Proposed D-TMSC algorithm – At the macrocell network

---

PHASE 1 – Initialization

• Each MUE $m \in \mathscr{L}_m$ initializes and broadcasts λ_m.

PHASE 2 – Lagrangian update

• Upon receiving all N "EXPIRE" packets, each MUE $m \in \mathscr{L}_m$ updates its λ_m by (6.17), broadcasts this value to all FUEs, resets its "EXPIRE" counter, and sets $t := t + 1$.

• Return to Phase 2 and repeat until convergence.

• *Between macrocell and cognitive femtocell networks*:

1. MUEs listen to "EXPIRE" packets from FUEs to decide when to update $\boldsymbol{\lambda}$.

2. MUEs broadcast the computed $\boldsymbol{\lambda}$ and the estimated $h_{k,m}^{(n)}$ to all FUEs.

6.3 A Dual Approach to Power-Efficient Resource Allocation

Recall that (6.9) aims at maximizing the rate-sum of the cognitive femtocell network, while specifying certain priority and/or fairness in sharing the secondary spectrum access opportunities among FUEs. In many other settings where FUEs are power-limited, it is even more important to achieve the power-efficient goal. Here, system throughput does not need to be enhanced at any cost. Rather, only a minimum data rate needs to be assured, giving a higher priority to power saving. These observations motivate us to formulate the following power-minimization problem.

$$\min_{\mathbf{p}} \ \sum_{k \in \mathscr{L}_f} w_k \sum_{n \in \mathscr{N}_k} p_k^{(n)} \tag{6.20a}$$

$$\text{s.t.} \ \sum_{k \in \mathscr{L}_f} \sum_{n \in \mathscr{N}_k} p_k^{(n)} \breve{I}_{k,m}^{(n)} \leq I_m^{\text{th}}, \quad \forall m \in \mathscr{L}_m \tag{6.20b}$$

$$\sum_{n \in \mathscr{N}_k} \ln \left(1 + \gamma_k^{(n)} p_k^{(n)} \right) \geq R_k^{\min}, \quad \forall k \in \mathscr{L}_f \tag{6.20c}$$

and (6.9d), (6.9e), (6.9f).

Unlike problem (6.9) which is always feasible, (6.20) is infeasible if all $R_k^{\min}$'s cannot be supported. Since (6.20) also satisfies the "frequency-sharing" condition, dual optimization can as well be employed to provide an optimal solution. The Lagrangian of (6.20) is defined as:

$$\mathscr{L}_e (\mathbf{p}, \mathbf{\nu}, \mathbf{\xi}) = \sum_{k \in \mathscr{L}_f} w_k \sum_{n \in \mathscr{N}_k} p_k^{(n)} + \sum_{m \in \mathscr{L}_m} \nu_m \left(\sum_{k \in \mathscr{L}_f} \sum_{n \in \mathscr{N}_k} p_k^{(n)} \breve{I}_{k,m}^{(n)} - I_m^{\text{th}} \right)$$
$$- \sum_{k \in \mathscr{L}_f} \xi_k \left(\sum_{n \in \mathscr{N}_k} \ln \left(1 + \gamma_k^{(n)} p_k^{(n)} \right) - R_k^{\min} \right), \tag{6.21}$$

with $\mathbf{\nu} = [\nu_1, \ldots, \nu_M]^T, \mathbf{\xi} = [\xi_1, \ldots, \xi_K]^T \succeq \mathbf{0}$. The Lagrange dual function is:

$$\bar{D}(\mathbf{\nu}, \mathbf{\xi}) = \sum_{n \in \mathscr{N}} \bar{D}^{(n)}(\mathbf{\nu}, \mathbf{\xi}) - \sum_{m \in \mathscr{L}_m} \nu_m I_m^{\text{th}} + \sum_{k \in \mathscr{L}_f} \xi_k R_k^{\min}, \tag{6.22}$$

with the per-subchannel problem being now:

$$\bar{D}^{(n)}(\mathbf{\nu}, \mathbf{\xi}) = \min_{\mathbf{p}} \sum_{k \in \mathscr{L}_f} \left\{ \left(w_k + \sum_{m \in \mathscr{L}_m} \nu_m \breve{I}_{k,m}^{(n)} \right) p_k^{(n)} - \xi_k \ln \left(1 + \gamma_k^{(n)} p_k^{(n)} \right) \right\}, \tag{6.23}$$

for all $n \in \mathscr{N}$.

Applying the KKT conditions and noting the exclusive channel assignment in OFDMA, we arrive at the optimal power allocation for active FUE $k \in \mathscr{L}_f$ on subchannel $n \in \mathscr{N}$ as:

$$
p_k^{(n)*} = \left(\frac{\xi_k}{w_k + \sum_{m \in \mathscr{L}_m} v_m \check{I}_{k,m}^{(n)}} - \frac{1}{\gamma_k^{(n)}} \right)^+ .
\tag{6.24}
$$

Since there are K FUEs in total, this solution gives rise to K possible values of $\bar{D}^{(n)}(\boldsymbol{v}, \boldsymbol{\xi})$ as:

$$
\bar{D}_k^{(n)}(\boldsymbol{v}, \boldsymbol{\xi}) = \left(w_k + \sum_{m \in \mathscr{L}_m} v_m \check{I}_{k,m}^{(n)} \right) p_k^{(n)*} - \xi_k \ln \left(1 + \gamma_k^{(n)} p_k^{(n)*} \right), \ \forall k \in \mathscr{L}_f.
\tag{6.25}
$$

To meet the spectrum-sharing constraints (6.9f), a procedure similar to that outlined in Table 6.1 can be used to provide an optimal subchannel-power allocation for each per-subchannel optimization problem. A distributed implementation can also be realized, similar to the throughput maximization scenario in Sect. 6.2. However, because the per-subchannel problems in this case involve the minimization operation, the virtual timer needs to be redefined as:

$$
\bar{T}_k^{(n)} = c_T \exp \left[\bar{D}_k^{(n)}(\boldsymbol{v}, \boldsymbol{\xi}) \right], \ k \in \mathscr{L}_f, n \in \mathscr{N}.
\tag{6.26}
$$

The overall dual function $\bar{D}(\boldsymbol{v}, \boldsymbol{\xi})$ in (6.22) can now be evaluated for the fixed $(\boldsymbol{v}, \boldsymbol{\xi})$. Finally, to solve the dual-domain problem

$$
\max_{\boldsymbol{v} \geq 0, \ \boldsymbol{\xi} \geq 0} \bar{D}(\boldsymbol{v}, \boldsymbol{\xi}),
\tag{6.27}
$$

the subgradient method can be employed as:

$$
v_m[t+1] = \left(v_m[t] - \delta_v \left[I_m^{\text{th}} - \sum_{k \in \mathscr{L}_f} \sum_{n \in \mathscr{N}_k} p_k^{(n)*} \check{I}_{k,m}^{(n)} \right] \right)^+, \quad m \in \mathscr{L}_m
\tag{6.28}
$$

$$
\xi_k[t+1] = \left(\xi_k[t] - \delta_\xi \left[\sum_{n \in \mathscr{N}_k} \ln \left(1 + \gamma_k^{(n)} p_k^{(n)*} \right) - R_k^{\min} \right] \right)^+, \quad k \in \mathscr{L}_f
\tag{6.29}
$$

with $\delta_v, \delta_\xi > 0$ being sequences of scalar step sizes.

Note that problem (6.20) can be infeasible due to the constraints on minimum attainable rates in (6.20c). In such cases, admission control should be performed to guarantee its feasibility. Among all FUEs whose throughput being below their respective threshold, we propose to drop the FUE with the largest deviation $|r_k - R_k^{\min}|$, and relinquish radio spectrum to the remaining FUEs. This procedure is repeated in at most K competing rounds[3] until all admitted FUEs have their minimum QoS requirements satisfied.

In this work, the resulting decentralized scheme is referred to as Distributed Energy-Efficient with Spectrum-sharing Constraints (D-EESC) algorithm. It is based on the D-TMSC design with the following modifications.

- *In the cognitive femtocell network*:

    1. Optimal power allocation for the per-subchannel problems is computed according to (6.24), instead of (6.14).
    2. Virtual timers at FUEs are defined in (6.26) with the optimal value of per-subchannel problem being now $\bar{D}_k^{(n)}(\nu, \xi)$ [see (6.25)].
    3. Each FUE $k \in \mathscr{L}_f$ updates ξ_k by (6.29), instead of (6.18).
    4. *Admission control*: At the end of each competing round, if all FUEs' $R_k^{\min}$ are satisfied, the algorithm terminates. Otherwise, the FUEs with $r_k^{\text{dev}} = r_k - R_k^{\min} < 0$ broadcast/exchange the information regarding their rate deviations. The FUE with the largest $|r_k^{\text{dev}}|$ will then voluntarily give up in all subsequent rounds and notify other FUEs and MUEs about this fact. The algorithm restarts and repeats until all r_k^{dev}'s are non-negative.

- *In the macrocell network*: MUEs update their cost variables ν according to (6.28), similar to (6.17).

6.4   Reduced-Complexity Schemes for Throughput Maximization

For the purpose of comparison, we develop two reduced-complexity resource allocation schemes to solve the throughput maximization problem (6.9). First, we propose a worst-case conservative design, which aims to satisfy the optimization constraints by applying the following two-stage procedure:

1. *Subchannel assignment*: The N available OFDM subchannels are allocated (in blocks) to individual FUEs to meet their minimum requirements on the number of subchannels $N_k^{\min}$. The remaining subchannels are then assigned in a random fashion to fully occupy the available radio spectrum. At this stage, constraints (6.9d)–(6.9f) are readily satisfied.

---

[3]One "competing round" means the time duration it takes for the algorithm to converge.

2. *Power allocation*: After subchannels have been designated to different FUEs, power needs to be distributed on those subchannels such that the constraints on interference (6.9b) and on total power (6.9c) are not violated. For (6.9b), upon noting that there are at most N subchannels to be occupied, it is clear that $\sum_{k \in \mathscr{L}_f} \sum_{n \in \mathscr{N}_k} p_k^{(n)} \check{I}_{k,m}^{(n)} \leq p_k^{(n)} (N \check{I}_m^{max})$, where $\check{I}_m^{max} = \max_{k,n} \check{I}_{k,m}^{(n)}$. Thus, a conservative design is to use $p_k^{(n)} = \min_m \{I_m^{th} / (N \check{I}_m^{max})\}$. To satisfy (6.9c), one may consider to equally split the power budget to the subchannels designated for each FUE, i.e., $p_k^{(n)} = P_k^{max} / |\mathscr{N}_k|$. Finally, the suggested heuristic power allocation is $\min \left\{ \min_m \{I_m^{th} / (N \check{I}_m^{max})\}, P_k^{max} / |\mathscr{N}_k| \right\}$.

Although the power allocation in the above worst-case design is simple, its conservativeness would likely offer a low system throughput. We propose another two-stage scheme that gives better performance with an affordable complexity. Specifically, the first stage of this heuristic design is identical to that of the worst-case solution. Once all the subchannels have been assigned to FUEs, it remains to solve the convex power allocation problem over those subchannels. For this, convex optimization softwares (e.g., CVX [25]) can be utilized to find the global optimum. Because of the separate optimization of subchannel assignment and power allocation, this scheme too can only give suboptimal performance.

6.5 Performance Evaluation

6.5.1 Asymptotic Complexity Analysis

Assume that searching through an unsorted 1-D list of dimension M requires a worst-case complexity of $\mathcal{O}(M)$. To resolve (6.9), the proposed centralized scheme needs to compute $K \times N$ matrix $\mathbf{A}$, entailing KN operations. For a fixed $\{\lambda, \mu\}$, this scheme involves N 2-D searches over $\mathbf{A}$ and thus demands $\mathcal{O}(N(KN)) = \mathcal{O}(KN^2)$ operations to find the solution of (6.12). Suppose that the subgradient method used to update $\{\lambda, \mu\}$ in (6.17) and (6.18) converges after τ iterations. It is reported in [21,22] that τ is a polynomial function of N. Computational experience also suggests that τ is relatively small for appropriate choices of step sizes δ_λ and δ_μ. On the other hand, the communication overheads incurred by the centralized design consist of the collection of network information to compute $\gamma_k^{(n)}$ and $\check{I}_{k,m}^{(n)}$. While constituting a complexity of $\mathcal{O}(MKN)$, this needs to be done only once for the whole allocation process. The complexity of the centralized scheme then totals to

$$\mathcal{O}\bigg(\underbrace{(KN + KN^2)\tau}_{\text{CMP}} + \underbrace{MKN}_{\text{OVH}} \bigg) = \mathcal{O}(KN^2\tau + MKN), \qquad (6.30)$$

where CMP and OVH denote computing efforts and communication overheads, respectively.

For the purpose of efficient processing, individual FUEs in the distributed scheme D-TMSC may sort their list of virtual timers in a decreasing order. This calls for $\mathcal{O}(N \log N)$ operations. During every one of the N competing minislots, each FUE searches through its own list to find and remove the subchannel that has already been used, implying a $\mathcal{O}(N)$ complexity. Hence, the number of operations needed for computing efforts at each FUE is $\mathcal{O}((N \log N + N)\tau) = \mathcal{O}(N \log(N)\tau)$. In contrast, for MUE it is simply $\mathcal{O}(\tau)$ since there is merely one update of λ_m per iteration.

The proposed distributed algorithm D-TMSC requires a very small amount of signalling information per iteration. From Table 6.2, each femto-Tx $k \in \mathcal{L}_f$ in Phase 1 needs to send only one flag message that contains the value of $N_k^{\min}$. In Phases 2 and 3, there are at most N "EXPIRE" massages to be broadcast by any single FUE. Similarly, each MUE $m \in \mathcal{L}_m$ is required to broadcast its computed Lagrangian λ_m only once in each iteration of the algorithm. As seen from Table 6.3, KN estimated values of $h_{k,m}^{(n)}$ are to be fed back from MUE $m \in \mathcal{L}_m$ to all FUEs $k \in \mathcal{L}_f$ only once per allocation period. In total, the D-TMSC algorithm entails an asymptotic complexity of

$$\mathcal{O}\Big(\underbrace{N \log(N)\tau}_{\text{CMP}} + \underbrace{1 + N\tau}_{\text{OVH}}\Big) = \mathcal{O}(N \log(N)\tau) \tag{6.31}$$

at individual femto-Tx's, and

$$\mathcal{O}(\underbrace{\tau}_{\text{CMP}} + \underbrace{KN}_{\text{OVH}}) \tag{6.32}$$

at individual MUEs.

On the other hand, admission control might be needed in the power-efficient scheme to ensure the feasibility of problem (6.20). While there are at most K competing rounds in this case, the computational complexity of each of which is similar to that in the throughput maximization case. It can be shown that the total worst-case complexity for the centralized power minimization scheme is

$$\mathcal{O}\left(K(KN^2\tau + MKN)\right) = \mathcal{O}(K^2N^2\tau + MK^2N), \tag{6.33}$$

whereas in the best case (i.e., without admission control) it is simply

$$\mathcal{O}(KN^2\tau + MKN). \tag{6.34}$$

By similar arguments, we can also conclude that the worst-case complexity (including all communication overheads) of the distributed D-EESC algorithm is $\mathcal{O}(KN \log(N)\tau)$ at each femto-Tx, and $\mathcal{O}(K\tau + K^2N)$ at each MUE. If admission control is not needed, the best-case complexity of D-EESC scheme is identical to that of D-TMSC.

Table 6.4 Asymptotic complexity analysis

Scheme		Total complexity	At each FUE	At each MUE
Rate Maximum	Optimal direct	$\mathcal{O}(K^N)$	–	–
	Centralized	$\mathcal{O}(KN^2\tau + MKN)$	–	–
	D-TMSC	–	$\mathcal{O}(N\log(N)\tau)$	$\mathcal{O}(\tau + KN)$
Power Minimum	Optimal direct	$\mathcal{O}(K^N)$	–	–
	Centralized	$\mathcal{O}(K^2N^2\tau + MK^2N)$	–	–
	D-EESC	–	$\mathcal{O}(KN\log(N)\tau)$	$\mathcal{O}(K\tau + K^2N)$

The foregoing asymptotic complexity analysis is summarized in Table 6.4. It is likely that direct approaches to solve (6.9) and (6.20) in the primal domain involve finding an optimal power allocation for every possible subchannel assignment. The latter operation alone imposes an exponential complexity, making these exhaustive-search resolutions computationally intractable for practical OFDM-based systems. On the contrary, the complexities of the two newly devised schemes are shown to only grow *polynomially* in the number of subchannels. This represents a substantial reduction in the complexity burden.

6.5.2 Illustrative Results

Consider a wireless communication scenario, in which a macrocell BS transmits downlink data to its $M = 2$ MUEs over predetermined frequencies in the available spectrum. All the macrocell signals are assumed to be elliptically filtered white noise with equal $\Phi_m(e^{jw}) = 1$. The frequency bands left unused by macrocell network are filled with N OFDM subchannels, over which $K = 3$ cognitive femtocell Tx-Rx pairs are allowed to communicate to exploit opportunistic spectrum access. In each simulation run, independent channel gains are randomly generated according to the Rayleigh distribution. The average channel gains, N_0 and B_N are all normalized to 1. We further assume that $w_1 = w_2 = w_3 = 1/3$.

6.5.2.1 Example 1: $N = 8$, $N^{\max} = [8, 8, 8]^T$, $N^{\min} = [0, 1, 2]^T$

This example aims to compare the performance of our proposed solutions with that of the globally optimal exhaustive search. The total number of accessible subchannels is limited to 8. The global optimum can be found by examining all possibilities of subchannels assignments, followed by solving the associated convex power-allocation problems. With $N = 8$ subchannels available, the total number of cases to be investigated by this direct method is $K^N = 3^8 = 6,561$, assuming every single subchannel is given to some FUE.

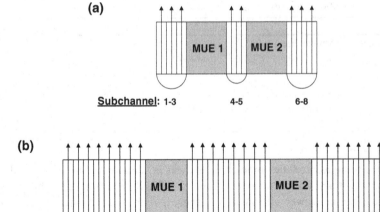

Fig. 6.2 Allocation of radio frequencies in the numerical examples. (a) Example 1 (b) Example 2

Figure 6.2a depicts the distribution of radio spectrum in Example 1. According to (6.1) and (6.5), the mutual interference between macrocell and cognitive femtocell networks depend on the spectral distances $\bar{d}_m^{(n)}$. We generate the normalized interferences from MUE 1 to the 8 OFDM subchannels as

$$\mathbf{J}_1 = [0.0678, 0.1525, 0.2712, 0.2712, 0.1525, 0.0678, 0.0169, 0]^T, \qquad (6.35)$$

and those from MUE 2 to all OFDM subchannels as $\mathbf{J}_2 = \mathbf{J}_1(8 : -1 : 1)$. For simplicity, we assume that the interferences from signals transmitted on these subchannels (excluding the effect of channel variations and power allocation) to MUE 1 and MUE 2 are $\hat{\mathbf{I}}_1 = \mathbf{J}_1$ and $\hat{\mathbf{I}}_2 = \mathbf{J}_2$, respectively. It is then possible to compute the mutual interferences $\check{I}_{k,m}^{(n)}$, $\check{J}_k^{(n)}$, and the CINR of individual FUEs $\gamma_k^{(n)}$.

As can be seen from Fig. 6.3a and b, the performances of the proposed dual schemes are almost indistinguishable with those of the direct exhaustive search. Note that in Fig. 6.3b, the values of minimum throughput $\mathbf{R}^{\min}$ have been selected such that problem (6.20) is always feasible. While the results shown in these two plots are consistent with the theory in [21, 22], these negligible duality gaps are actually attained with only 8 subchannels. As the number of OFDM subchannels increases, the difference between primal optimal value and its dual counterpart is expected to become even smaller. We also exhibit in Fig. 6.3c, d the convergence process of both D-TMSC and D-EESC schemes. In all simulations, we have used the diminishing step size rule $\delta[t] = 1/\sqrt{t}$ and tolerance $\epsilon = 10^{-4}$ for these two algorithms. As seen, the devised schemes take merely around 100 iterations to quickly converge to stable solutions.

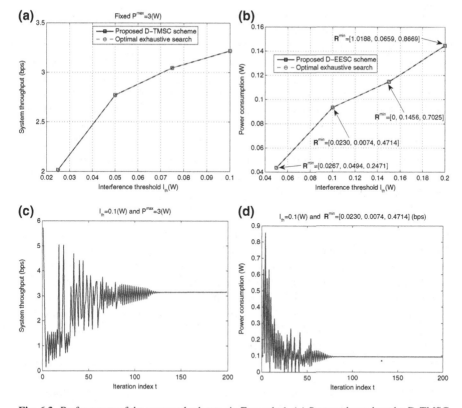

Fig. 6.3 Performance of the proposed schemes in Example 1. (**a**) System throughput by D-TMSC scheme. (**b**) Power consumption by D-EESC scheme. (**c**) Convergence of D-TMSC scheme. (**d**) Convergence of D-EESC scheme

6.5.2.2   Example 2: $N = 24, \mathbf{N}^{\text{max}} = [12, 24, 12]^T, \mathbf{N}^{\text{min}} = [4, 8, 4]^T$

This example assumes $N = 24$ subchannels available for use by the femtocell network, whose specific locations in the frequency domain are shown in Fig. 6.2b. Note that it is already computationally prohibited to perform exhaustive search in this case. Similar to Example 1, we generate the mutual interferences as follows:

$$\mathbf{J}_1 = \big[0.0274, 0.0346, 0.0427, 0.0517, 0.0615, 0.0722, 0.0838, 0.0962, 0.0962,$$

$$0.0838, 0.0722, 0.0615, 0.0517, 0.0427, 0.0346, 0.0274, 0.0209, 0.0154,$$

$$0.0107, 0.0068, 0.0038, 0.0017, 0.0004, 0.0000\big]^T,$$

$\mathbf{J}_2 = \mathbf{J}_1(24 : -1 : 1)$, $\hat{\mathbf{I}}_1 = \mathbf{J}_1$ and $\hat{\mathbf{I}}_2 = \mathbf{J}_2$. Here, the final result of each simulation run is averaged over 100 channel realizations.

We investigate the performance of our proposed method D-TMSC by comparing it with that of the worst-case conservative and the heuristic designs presented in Sect. 6.4. In these comparisons, perfect knowledge of network conditions is first assumed. While this assumption can be valid for many applications, we also evaluate the performance of D-TMSC when there is only imperfect network information available, possibly due to erroneous spectrum detection, user mobility, and limitation of channel estimation algorithms. As can be seen from (6.9), the quality of our proposed solution is affected ultimately because of the variations in $\check{I}_{k,m}^{(n)}$ and $\gamma_k^{(n)}$. Further investigations of (6.5) and (6.6) reveal that the robustness of the devised approach relies upon both the spectral dynamics of PUs (which results in varying B_m) and Doppler effect caused by geographical mobility of both FUEs and MUEs (which leads to varying $h_{l,l'}^{(n)}$, $l, l' \in \mathscr{L}, n \in \mathscr{N}$).

To examine this, we allow imperfect values $\bar{I}_{k,m}^{(n)} = \check{I}_{k,m}^{(n)}(1 + \Delta_I)$ and $\bar{\gamma}_k^{(n)} = \gamma_k^{(n)}(1 - \Delta_\gamma)$, where Δ_I and Δ_γ are random variables uniformly distributed in the interval $[0, \Delta_I^{\max}]$ and $[0, \Delta_\gamma^{\max}]$, respectively. In practice, the error thresholds $\Delta_I^{\max}, \Delta_\gamma^{\max}$ can be set based on (i) performance bound of the underlying spectrum-detection algorithms, and (ii) the maximum predictable speed of UEs. Figure 6.4a, b show the achieved throughput of all schemes for fixed values of $I^{\text{th}} = I_1^{\text{th}} = I_2^{\text{th}} = 0.05W$ and $P^{\max} = 3W$, respectively. It is clear from these figures that the distributed algorithm D-TMSC outperforms both conservative and heuristic designs. Interestingly, a significant performance advantage can still be realized even if the optimization at the cognitive femtocell network is based on imperfect network information.

Depicted in Fig. 6.4c is the total number of OFDM subchannels allotted to individual FUEs by the D-TMSC scheme, averaged over all channel realizations. It is straightforward to see that our proposed D-TMSC approach, in any case, guarantees to fulfill the desirable spectrum requirements of each FUE $k \in \mathscr{L}_f$, specified by $N_k^{\max}$ and $N_k^{\min}$. Yet, these subchannels are not equally utilized as some are distributed with more power than the others. This can be best observed in Fig. 6.4d, where most of the OFDM subchannels adjacent to the MUE spectrum (from subchannel 7 to 18) are shown to be allocated with very little power, as a result of being highly interfered by macrocell network signals.

For these subchannels, their corresponding $D_k^{(n)}$ entries are likely zero. In the case that all the remaining entries are zero, there are unnecessary ties in the competition among the FUEs. Here, we propose to designate these ineffectual channels to the FUEs in the order of priority, until all spectrum-sharing constraints are satisfied. In our simulation, FUE 1 and FUE 3 are given the highest and the lowest priority, respectively. Since these channels are useless anyway, the performance of the distributed scheme is not affected while a back-off mechanism is avoided. This also explains why FUE 1 appears to possess more subchannels than the other two FUEs in Fig. 6.4c.

To verify the performance of the proposed D-EESC scheme, the worst-case power allocation $p_k^{(n)} = \min_m \left\{ I_m^{\text{th}} / [N \check{I}_m^{\max}] \right\}, \forall k \in \mathscr{L}_f, n \in \mathscr{N}$ is used to provide a baseline. Also included in the comparison is the heuristic design that involves a fixed assignment of subchannels followed by an optimal power allocation, similar

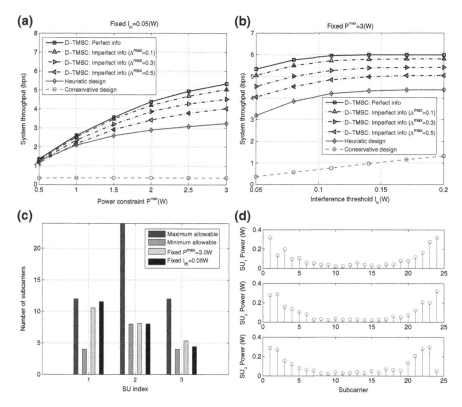

Fig. 6.4 Performance of D-TMSC scheme in Example 2. (**a**) Achieved throughput for fixed I^{th} (assuming $\Delta_I^{\max} = \Delta_\gamma^{\max} = \Delta^{\max}$) (**b**) Achieved throughput for fixed $P^{\max}$ (assuming $\Delta_I^{\max} = \Delta_\gamma^{\max} = \Delta^{\max}$) (**c**) Assignment of subchannels (averaged for $P^{\max} = 3.0W$ and $I^{\text{th}} = 0.05W$) (**d**) Allocation of power (averaged for $P^{\max} = 3.0W$ and $I^{\text{th}} = 0.05W$)

to that in the case of throughput maximization. Recall that problem (6.20) will become infeasible if all $R_k^{\min}$'s cannot be supported. To circumvent this issue, in our numerical examples the data rates attained by the worst-case solution are used to set the minimum required throughput of individual FUEs. Doing so will guarantee the feasibility of (6.20) and, as a consequence, admission control is no longer necessary.

Figure 6.5 displays the amounts of power consumed by all the considered schemes for a given I^{th}. Apparently, the proposed D-EESC algorithm requires considerably less amount of power to satisfy the minimum QoS compared to the other suboptimal designs. The improvement is even more pronounced at high values of interference threshold. For instance, at $I^{\text{th}} = 0.3W$ the saving in power is well-above 60% and 30% over the worst-case design and the heuristic solution, respectively. Furthermore, although the performance of D-EESC scheme in the case of imperfect network information is degraded, it is still noticeably better than those of the other two approaches.

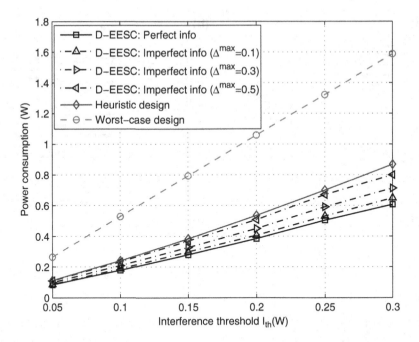

Fig. 6.5 Performance of D-EESC scheme in Example 2 (assuming $\Delta_I^{max} = \Delta_\gamma^{max} = \Delta^{max}$)

References

1. G. Gur, S. Bayhan, and F. Alagoz, "Cognitive femtocell networks: An overlay architecture for localized dynamic spectrum access [Dynamic Spectrum Management]," *IEEE Wirel. Commun.*, vol. 17, no. 4, pp. 62–70, 2010.
2. S. Al-Rubaye, A. Al-Dulaimi, and J. Cosmas, "Cognitive femtocell," *IEEE Veh. Technol. Mag.*, vol. 6, no. 1, pp. 44–51, 2011.
3. R. Xie, F. Yu, H. Ji, and Y. Li, "Energy-efficient resource allocation for heterogeneous cognitive radio networks with femtocells," *IEEE Trans. Wireless Commun.*, vol. 11, no. 11, pp. 3910–3920, 2012.
4. D. T. Ngo, S. Khakurel, and T. Le-Ngoc, "Distributed subchannel and power allocation for OFDMA-based femtocell networks," in *IEEE Vehicular Technology Conf. (VTC-Spring)*, Dresden, Germany, Jun. 2013, pp. 1–5.
5. D. T. Ngo, S. Khakurel, and T. Le-Ngoc, "Joint subchannel assignment and power allocation for OFDMA femtocell networks," *IEEE Trans. Wireless Commun.*, vol. 13, no. 1, pp. 342–355, Jan. 2014.
6. D. T. Ngo and T. Le-Ngoc, "Distributed resource allocation for cognitive radio ad-hoc networks with spectrum-sharing constraints," in *Proc. IEEE Global Commun. Conf. (GLOBECOM)*, Miami, FL, USA, Dec. 2010, pp. 1–6.
7. D. T. Ngo and T. Le-Ngoc, "Distributed resource allocation for cognitive radio networks with spectrum-sharing constraints," *IEEE Trans. Veh. Technol.*, vol. 60, no. 7, pp. 3436–3449, Sep. 2011.
8. T. Clancy, "Formalizing the interference temperature model," *Wirel. Commun. Mob. Comput. (UK)*, vol. 7, no. 9, pp. 1077–1086, 2007.

9. Y. Xing, C. N. Mathur, M. Haleem, R. Chandramouli, and K. Subbalakshmi, "Dynamic spectrum access with QoS and interference temperature constraints," *IEEE Trans. Mobile Computing*, vol. 6, no. 4, pp. 423–433, Apr. 2007.

10. W. Rhee and J. Cioffi, "Increasing in capacity of multiuser OFDM systems using dynamic subchannel allocation," in *Proc. IEEE Vehicular Technology Conf. (VTC)*, vol. 2, May 2000, pp. 1085–1089.

11. Z. Shen, J. G. Andrews, and B. L. Evans, "Adaptive resource allocation in multiuser OFDM systems with proportional rate constraints," *IEEE Trans. Wireless Commun.*, vol. 4, no. 6, pp. 2726–2737, Nov. 2005.

12. C. Bae and D.-H. Cho, "Fairness-aware adaptive resource allocation scheme in multihop OFDMA systems," *IEEE Commun. Lett.*, vol. 11, no. 2, pp. 134–136, Feb. 2007.

13. D. T. Ngo, C. Tellambura, and H. H. Nguyen, "Efficient resource allocation for OFDMA multicast systems with spectrum-sharing control," *IEEE Trans. Veh. Technol.*, vol. 58, no. 9, pp. 4878–4889, Nov. 2009.

14. Y. Otani, S. Ohno, K. Teo, and T. Hinamoto, "Subcarrier allocation for multi-user OFDM system," in *Proc. Asia-Pacific Conf. on Commun.*, Oct. 2005, pp. 1073–1077.

15. C. Suh and J. Mo, "Resource allocation for multicast services in multicarrier wireless communications," *IEEE Trans. Wireless Commun.*, vol. 7, no. 1, pp. 27–31, Jan. 2008.

16. P. Cheng, Z. Zhang, H.-H. Chen, and P. Qiu, "Optimal distributed joint frequency, rate and power allocation in cognitive OFDMA systems," *IET Communications*, vol. 2, no. 6, pp. 815–826, Jul. 2008.

17. Y. Ma, D. I. Kim, and Z. Wu, "Optimization of OFDMA-based cellular cognitive radio networks," *IEEE Trans. Commun.*, vol. 58, no. 8, pp. 2265–2276, Aug. 2010.

18. T. Weiss, J. Hillenbrand, A. Krohn, and F. Jondral, "Mutual interference in OFDM-based spectrum pooling systems," in *Proc. IEEE Vehicular Technology Conf. (VTC)*, vol. 4, May 2004, pp. 1873–1877.

19. Z. Hasan, G. Bansal, E. Hossain, and V. Bhargava, "Energy-efficient power allocation in OFDM-based cognitive radio systems: A risk-return model," *IEEE Trans. Wireless Commun.*, vol. 8, no. 12, pp. 6078–6088, Dec. 2009.

20. T. Rappaport, *Wireless Communications: Principles and Practice*, 2nd ed. Upper Saddle River, NJ, USA: Prentice Hall, 2001.

21. K. Seong, M. Mohseni, and J. M. Cioffi, "Optimal resource allocation for OFDMA downlink systems," in *Proc. IEEE Intl. Symp. on Inform. Theory*, Jul. 2006, pp. 1394–1398.

22. W. Yu and R. Lui, "Dual methods for nonconvex spectrum optimization of multicarrier systems," *IEEE Trans. Commun.*, vol. 54, no. 7, pp. 1310–1322, Jul. 2006.

23. D. P. Bertsekas, *Nonlinear Programming*, 2nd ed. Boston: Athena Scientific, 1999.

24. P802.11, "IEEE standard for wireless LAN medium access control (MAC) and physical layer (PHY) specifications," Nov. 1997.

25. M. Grant and S. Boyd, "CVX: Matlab software for disciplined convex programming, version 1.21," http://cvxr.com/cvx, Jan. 2011.